WINDOWS AZURE® MOBILE

INTRODUCTION		XV
CHAPTER 1	Introduction and Fundamental Concepts	1
CHAPTER 2	Creating and Manipulating Data	17
CHAPTER 3	Mobile Services Validation	39
CHAPTER 4	Authentication Options in WAMS	58
CHAPTER 5	Using REST to Access WAMS Data	79
CHAPTER 6	Push Notifications	93
CHAPTER 7	Advanced Scripting	107
CHAPTER 8	Advanced Configuration	126
INDEX		137

Windows Azure® Mobile Services

Windows Azure® Mobile Services

Bruce Johnson

A Wiley Brand

Windows Azure® Mobile Services

Published by
John Wiley & Sons, Inc.
10475 Crosspoint Boulevard
Indianapolis, IN 46256
www.wiley.com

Copyright © 2013 by John Wiley & Sons, Inc., Indianapolis, Indiana

ISBN: 978-1-118-67869-5
ISBN: 978-1-118-67853-4 (ebk)
ISBN: 978-1-118-74991-3 (ebk)

No part of this publication may be reproduced, stored in a retrieval system or transmitted in any form or by any means, electronic, mechanical, photocopying, recording, scanning or otherwise, except as permitted under Sections 107 or 108 of the 1976 United States Copyright Act, without either the prior written permission of the Publisher, or authorization through payment of the appropriate per-copy fee to the Copyright Clearance Center, 222 Rosewood Drive, Danvers, MA 01923, (978) 750-8400, fax (978) 646-8600. Requests to the Publisher for permission should be addressed to the Permissions Department, John Wiley & Sons, Inc., 111 River Street, Hoboken, NJ 07030, (201) 748-6011, fax (201) 748-6008, or online at http://www.wiley.com/go/permissions.

Limit of Liability/Disclaimer of Warranty: The publisher and the author make no representations or warranties with respect to the accuracy or completeness of the contents of this work and specifically disclaim all warranties, including without limitation warranties of fitness for a particular purpose. No warranty may be created or extended by sales or promotional materials. The advice and strategies contained herein may not be suitable for every situation. This work is sold with the understanding that the publisher is not engaged in rendering legal, accounting, or other professional services. If professional assistance is required, the services of a competent professional person should be sought. Neither the publisher nor the author shall be liable for damages arising herefrom. The fact that an organization or Web site is referred to in this work as a citation and/or a potential source of further information does not mean that the author or the publisher endorses the information the organization or Web site may provide or recommendations it may make. Further, readers should be aware that Internet Web sites listed in this work may have changed or disappeared between when this work was written and when it is read.

For general information on our other products and services please contact our Customer Care Department within the United States at (877) 762-2974, outside the United States at (317) 572-3993 or fax (317) 572-4002.

Trademarks: Wiley, the Wiley logo, Wrox, the Wrox logo, Programmer to Programmer, and related trade dress are trademarks or registered trademarks of John Wiley & Sons, Inc. and/or its affiliates, in the United States and other countries, and may not be used without written permission. Windows Azure is a registered trademark of Microsoft Corporation. All other trademarks are the property of their respective owners. John Wiley & Sons, Inc., is not associated with any product or vendor mentioned in this book.

ACQUISITIONS EDITOR
Mary James

PROJECT EDITOR
Edward Connor

TECHNICAL EDITOR
Wouter van Eck

PRODUCTION EDITOR
Christine Mugnolo

COPY EDITOR
Kim Cofer

EDITORIAL MANAGER
Mary Beth Wakefield

FREELANCE EDITORIAL MANAGER
Rosemarie Graham

ASSOCIATE DIRECTOR OF MARKETING
David Mayhew

MARKETING MANAGER
Ashley Zurcher

VICE PRESIDENT AND EXECUTIVE GROUP PUBLISHER
Richard Swadley

VICE PRESIDENT AND EXECUTIVE PUBLISHER
Neil Edde

ASSOCIATE PUBLISHER
Jim Minatel

PROOFREADER
James Saturnio, Word One New York

COVER DESIGNER
Ryan Sneed

I'd like to thank my wife, Ali, and my four children, Curtis, Gillian, Cameron, and Kyle, for their love and support. They patiently put up with my lack of time whenever I tackle a book project, and for that I am truly grateful. And this time around, I have to include Hershey in the dedication. So say my kids . . . and what they say goes ☺.

– *Bruce Johnson.*

ABOUT THE AUTHORS

BRUCE JOHNSON is a partner at ObjectSharp Consulting and a 30-year veteran of the computer industry. The first third of his career was spent doing "real work," otherwise known as coding in the UNIX world. But for almost 20 years, he has been working on projects that are at the leading edge of Windows technology, from C++ through Visual Basic to C#, and from thick client applications to websites to services.

As well as having fun with building systems, Bruce has spoken hundreds of times at conferences and user groups throughout North America. He has been a Microsoft Certified Trainer (MCT) for the past three years and he is a co-president of the Metro Toronto .NET User Group. He has written columns and articles for numerous magazines. While the quantity of the posts on his blog (http://www.objectsharp.com/blogs/bruce) have decreased recently, the activity on his Twitter account (http://www.twitter.com/lacanuck) has shown a corresponding increase. For all of this activity (or, perhaps, in spite of it), Bruce has been privileged to be recognized as a Microsoft MVP for the past six years.

ABOUT THE TECHNICAL EDITOR

WOUTER VAN ECK is a software design and development consultant specializing in cloud integration with a strong focus on the Microsoft platform. In the past five years or so, Wouter has focused on what we now call "the cloud" as the base for integration work backed by the Microsoft Azure technology. With successful deployments in Canada, the United States, Germany, and Singapore, Wouter has demonstrated true cloud experience.

ACKNOWLEDGMENTS

THE WRITING OF A BOOK is much more of a collaborative effort than might be apparent to an outsider. I have written a number of books and I can guarantee you that there is no way a single one of those books would have seen the light of day without a great deal of assistance from many other people. There are a surprising number of people involved in the editorial process, and odds are pretty good that if you can understand the content of the book, that my editor, my technical reviewer, and my copy editor are the reason why. My first draft is not something that should be read by anyone, and these people shape the pages so that the result is something that can be read by everyone. That is not an easy task, and I am indebted to them for their assistance.

I would especially like to thank everyone at Wrox who has helped me through this process. In particular, thanks go out to Ed Connor, whose patience and attention to detail is quite impressive. Thanks also go to Wouter van Eck, who did a great job making sure that the technical details of the book were accurate. Finally, thanks to Kim Cofer, who had the unenviable chore of ensuring that I wasn't writing in the passive voice, and fixing it when I stopped writing so well. The efforts of all of these individuals are what make the book possible and, hopefully, a success. Thanks also to Mary James, who was kind enough to allow me to indulge my passion to write.

Lastly, I would like to thank all of my associates at ObjectSharp and the people at Microsoft who, although they might not have realized it, were keeping the writing process going by answering any questions I had.

— Bruce Johnson

CONTENTS

INTRODUCTION TO THE WINDOWS AZURE BOOK SERIES	*xv*
INTRODUCTION TO WINDOWS AZURE MOBILE SERVICES	*xvii*
CHAPTER 1: INTRODUCTION AND FUNDAMENTAL CONCEPTS	**1**
What Is Azure?	2
A Brief History of Azure	2
Windows Azure Features	3
Execution Model	3
Data Storage	5
Connectivity	6
Authentication	7
Messaging	8
And Now, Back to Our Show	8
Setting Up WAMS	9
Generating a Sample Application	13
Summary	16
CHAPTER 2: CREATING AND MANIPULATING DATA	**17**
The Data Model	18
Client-Side Functionality	26
Summary	38
CHAPTER 3: MOBILE SERVICES VALIDATION	**39**
Adding Server-Side Scripts	39
Inserting Data	42
Updating Data	43
Deleting Data	44
Retrieving Data	45
The User Object	47
Common Scenarios	48
Summary	57

CONTENTS

CHAPTER 4: AUTHENTICATION OPTIONS IN WAMS — 58

- Federated Authentication — 58
- Setting Up the Authentication Providers — 61
- Configuring Your Service for Authentication — 72
- Authentication on the Client Side — 74
- Troubleshooting Authentication — 77
- **Summary** — 78

CHAPTER 5: USING REST TO ACCESS WAMS DATA — 79

- **Representational State Transfer** — 79
  - GET — 80
  - POST — 80
  - PUT — 81
  - DELETE — 81
- **REST and WAMS** — 82
  - Authentication in REST — 88
- **Summary** — 92

CHAPTER 6: PUSH NOTIFICATIONS — 93

- **Registering Your Application** — 95
  - Windows Notification Services — 95
  - Google Cloud Messaging — 98
  - Apple Push Notification Services — 99
- **Configuring Your Mobile Service** — 100
- **The Mechanics of Push Notifications** — 101
  - Requesting a Channel — 101
- **Sending Notifications** — 104
- **Summary** — 106

CHAPTER 7: ADVANCED SCRIPTING — 107

- **Auditing Updates** — 107
- **Supporting Other Data Types** — 109
- **Supporting Arrays** — 113
- **Additional User Information** — 118
  - Facebook — 118
  - Google — 119
  - Microsoft Account — 120
  - Twitter — 120
- **Scheduling Tasks** — 121
- **Summary** — 125

CHAPTER 8: ADVANCED CONFIGURATION 126

Scaling WAMS 126
Horizontal versus Vertical Scaling? 127
Using an Existing Database 130
Monitoring WAMS 133
Summary 136

INDEX *137*

INTRODUCTION TO THE WINDOWS AZURE BOOK SERIES

It has been fascinating watching the maturation of Windows Azure since its introduction in 2008. When it was announced, Azure was touted as being Microsoft's "new operating system." And at that level, it has not really lived up to its billing. However, if you consider Azure to be a collection of platforms and tools that allow you to cloud-enable your corporation's applications and infrastructure, well, now you're on the right track.

And, as it turns out, a collection of co-operating tools and services is the best way to think of Azure. The different components that comprise Azure become building blocks that allow you to construct an environment to suit your needs. Want to be able to host a simple website? Well, then Azure Web Sites fits the bill. Want to move some of your infrastructure to the cloud while leaving other systems on-premises? Azure Virtual Networking gives you the capability to extend your corporate domain to include machines hosted in Azure. Almost without exception, each twist and turn in your infrastructure roadmap can take advantage of the building blocks that make up Windows Azure.

A single book covering everything that encompasses Azure would be huge. And because of the breadth of components in Azure, such a book is likely to contain information that you are not necessary interested in. For this reason, the Windows Azure series from Wrox takes the same "building block" approach that Azure does. Each book in the series drills deeply into one technology. If you want to learn everything you need to work with a particular technology, then you could not do better than to pick up the book for that topic. But you don't have to dig through 2,000 pages to find the 120 pages that matter to you. Each book stands on its own. You can pick up the books for the topics you are care about and know that's all that you will get. And you can leave the other books until desire or circumstance makes them of interest to you.

So enjoy this book. It will give you the information you need to put Windows Azure to use for you. But as you continue to look to other Azure components to add to your infrastructure, don't forget to check out the other books in the series to see what topics might be helpful. The books in the series are:

- *Windows Azure and ASP.NET MVC Migration* by Benjamin Perkins, Senior Support Escalation Engineer, Microsoft
- *Windows Azure Mobile Services* by Bruce Johnson, MVP, Partner, ObjectSharp Consulting
- *Windows Azure Web Sites* by James Chambers, Product & Community Development Manager, LogiSense

- *Windows Azure Data Storage* Simon Hart, Software Architect, Microsoft
- *Windows Azure Hybrid Cloud* by Danny Garber, Windows Azure Solution Architect, Microsoft; Jamal Malik, Business Solution Architect; Adam Fazio, Solution Architect, Microsoft

Each one of these books was written with the same thought in mind: to provide deep knowledge of that one topic. As you go further into Azure, you can pick and choose what makes sense for you from the other books that are available. Constructing your knowledge using these books is like building blocks. Which is just the same manner that Azure was designed.

Bruce Johnson
Azure Book Series Editor

INTRODUCTION TO *WINDOWS AZURE MOBILE SERVICES*

WINDOWS AZURE has had a brief, yet interesting history. Originally pitched as Microsoft's new operating system, it was launched to great fanfare. Though its acceptance was not initially overwhelming, the functionality that it offered was of interest in certain areas of IT operation and development. But, more importantly, it continued to grow. Over time, more and more features were added and Microsoft put the infrastructure in a place necessary to support these features.

June 2012 was the real turning point for Windows Azure. A major announcement introduced a number of new and very compelling features. From websites to virtual machines to hybrid networking solutions, the reasons for people to pay attention to Windows Azure increased significantly. And from that moment, Microsoft has not slowed down the pace.

The biggest challenge faced by Windows Azure is now recognition. Many developers are either unaware of the features that are provided, or still relate to it as solely the "operating system" as which it was first released. The purpose of this book (and indeed, the purpose of the Windows Azure series of books that is published by Wiley) is to change that perception. By presenting a soup-to-nuts look at what Windows Azure Mobile Services (WAMS) has to offer, the goal is to provide you with enough information to either start using it immediately, or to put it into your development toolbox.

WHO THIS BOOK IS FOR

Windows Azure Mobile Services is for any developer who is new to Windows Azure as well as those programmers who have some experience with Azure, but might have only dabbled in the features as they came. The biggest problem with Azure Mobile Services is not a lack of functionality, but a lack of familiarity. A high percentage of developers have little or no experience with Azure, but when you start to talk about the features, their eyes light up. WAMS is frequently a source for this reaction.

If you're just starting out with WAMS, you'll benefit greatly from the first two chapters, where the initial configuration and basic data functionality is covered. If you have some experience with Mobile Services, these chapters might be a refresher, but the book quickly moves into more advanced topics, including REST-based access, the details of server-side scripting, and advanced configuration scenarios. Regardless of your previous experience, this book has something useful to offer.

WHAT THIS BOOK COVERS

This book covers the complete spectrum of WAMS functionality. It starts with a discussion of Windows Azure and how to set up your trial account, define your first mobile service, and create a simple demonstration application. It then moves into the functions that are at the heart of WAMS: data creation, modification, and retrieval.

Beyond these fundamentals, the book goes into more and more detail about the capabilities of WAMS. This includes the ability to verify data using server-side script, the different options that are available when it comes to accessing the data, and the scalability and manageability of the service itself. The topics allow a novice user to get from ground zero to being able to take full advantage of WAMS, and to expose more experienced developers to areas that they might not already be familiar with.

HOW THIS BOOK IS STRUCTURED

This book is structured with the aim of providing start-to-finish guidance to a person who has no previous experience with Windows Azure Mobile Services. To this end, the early chapters cover the fundamentals of setting up WAMS, accessing the data from a client application, and creating server-side functionality. As you progress through the book, the topics become more advanced and more complex, while still remaining firmly rooted in the real world. By the end of the book (presuming you read it in order), you will be well positioned to utilize WAMS for almost any use.

WHAT YOU NEED TO USE THIS BOOK

To use this book effectively, you'll need only a few pieces of software. To start with, the book assumes that you have access to Microsoft Visual Studio 2012. The edition of Visual Studio that you use does not matter too much. There might be some types of projects that you can't open with the Express Edition, but the SDK that is used with Windows Azure Mobile Services will work in any addition.

A number of different SDKs are available to assist with creating the client-side of any application that wants to utilize WAMS. In fact, there is a mobile SDK for each of the four platforms (Windows Store, Windows Phone, iOS, and Android) that are supported. It is not required that you use an SDK in order to work with your mobile service. However, if you are creating applications using any of those platforms, you might find the classes provided by the SDK useful. You can download the SDKs from `http://www.windowsazure.com/en-us/downloads`.

Along with these tools, you also need to have a Windows Azure account in order to create your mobile service. You can find the necessary steps to get a trial account (presuming you don't already have one) in Chapter 1.

The source code for the samples is available for download from the Wrox website at:

    www.wrox.com/go/windowsazuremobileservices

CONVENTIONS

To help you get the most from the text and keep track of what's happening, we've used a number of conventions throughout the book.

> **NOTE** *Notes indicate notes, tips, hints, tricks, and/or asides to the current discussion.*

As for styles in the text:

- We *highlight* new terms and important words when we introduce them.
- We show keyboard strokes like this: Ctrl+A.
- We show filenames, URLs, and code within the text like so: `persistence.properties`.
- We use a monofont type with no highlighting for most code examples.

SOURCE CODE

As you work through the examples in this book, you may choose either to type in all the code manually, or to use the source code files that accompany the book. All the source code used in this book is available for download at `www.wrox.com`. Specifically for this book, the code download is on the Download Code tab at:

    www.wrox.com/go/windowsazuremobileservices

You can also search for the book at `www.wrox.com` by ISBN (the ISBN for this book is 9781118678534 to find the code. And a complete list of code downloads for all current Wrox books is available at `www.wrox.com/dynamic/books/download.aspx`.

Most of the code on `www.wrox.com` is compressed in a ZIP, RAR archive, or similar archive format appropriate to the platform. Once you download the code, just decompress it with an appropriate compression tool.

> **NOTE** *Because many books have similar titles, you may find it easiest to search by ISBN; this book's ISBN is 978-111-867853-4.*

Once you download the code, just decompress it with your favorite compression tool. Alternatively, you can go to the main Wrox code download page at `www.wrox.com/dynamic/books/download.aspx` to see the code available for this book and all other Wrox books.

ERRATA

We make every effort to ensure that there are no errors in the text or in the code. However, no one is perfect, and mistakes do occur. If you find an error in one of our books, like a spelling mistake or faulty piece of code, we would be very grateful for your feedback. By sending in errata, you may save another reader hours of frustration, and at the same time, you will be helping us provide even higher-quality information.

To find the errata page for this book, go to:

> `www.wrox.com/go/windowsazuremobileservices`

and click the Errata link. On this page you can view all errata that has been submitted for this book and posted by Wrox editors.

If you don't spot "your" error on the Book Errata page, go to `www.wrox.com/contact/techsupport.shtml` and complete the form there to send us the error you have found. We'll check the information and, if appropriate, post a message to the book's errata page and fix the problem in subsequent editions of the book.

P2P.WROX.COM

For author and peer discussion, join the P2P forums at `http://p2p.wrox.com`. The forums are a web-based system for you to post messages relating to Wrox books and related technologies and interact with other readers and technology users. The forums offer a subscription feature to e-mail you topics of interest of your choosing when new posts are made to the forums. Wrox authors, editors, other industry experts, and your fellow readers are present on these forums.

At `http://p2p.wrox.com`, you will find a number of different forums that will help you, not only as you read this book, but also as you develop your own applications. To join the forums, just follow these steps:

1. Go to `http://p2p.wrox.com` and click the Register link.
2. Read the terms of use and click Agree.
3. Complete the required information to join, as well as any optional information you wish to provide, and click Submit.
4. You will receive an e-mail with information describing how to verify your account and complete the joining process.

> **NOTE** *You can read messages in the forums without joining P2P, but in order to post your own messages, you must join.*

Once you join, you can post new messages and respond to messages other users post. You can read messages at any time on the web. If you would like to have new messages from a particular forum e-mailed to you, click the Subscribe to this Forum icon by the forum name in the forum listing.

For more information about how to use the Wrox P2P, be sure to read the P2P FAQs for answers to questions about how the forum software works, as well as many common questions specific to P2P and Wrox books. To read the FAQs, click the FAQ link on any P2P page.

happy, but it allowed for dedicated virtual machines to be included in a deployment (that's the VM role portion) and, more interestingly, allowed that VM to connect securely to resources that were inside a corporate network. For example, a web application could now access a local SQL resource without using a Windows Communication Foundation (WCF) interface or exposing the SQL database directly to the Internet.

The next few releases of Azure were also incremental in nature. They extended and solidified the functionality that was already in place. But there really wasn't anything to make the general developer population sit up and take notice. That changed with the June 2012 release.

Even to people who had been involved with Windows Azure for a while, the June 2012 release was a revelation. The list of features, more or less, mirrors the list of features that are covered in the next section — Azure Web Sites, Virtual Machines, and support for non-proprietary environments, such as Linux. This release has a richness of functionality that dwarfs what came before. More importantly, it is no longer a requirement to re-architect an existing web application to take advantage of Azure. You could deploy your existing applications, in some cases with no change at all, to Azure using MSDeploy. Developers really sat up and took notice.

The pricing model was quite appealing. It made Azure competitive with other hosting companies. As well, Microsoft added a spending limit on the monthly cost (set by default to $0) ensuring that you wouldn't get charged an unexpectedly large sum of money at the end of the billing period. Over the rest of this chapter you take a look at the current offerings from Azure (with the emphasis on "current" as of this writing...odds are new features will be added before this book goes to print).

WINDOWS AZURE FEATURES

Azure functionality can be aligned across two dimensions. First, there is the "service-based" delineation: Infrastructure as a Service (IaaS), Platform as a Service (PaaS), and Software as a Service (SaaS). These distinctions align, more or less, to the level of control that you (as the purchaser of the service) have over the hardware and operating system on which the service is running. However, this is not (at least for the expected audience for this book) a particularly useful breakdown. In other words, I could simply list out the features that Windows Azure has, but that's not going to be particularly useful to you, the reader, as you try to figure out what pieces of Azure you may or may not care about. So instead, I divide the features of Azure into functional chunks that more directly relate to how you are likely to use them.

Execution Model

Regardless of how you look at it, one of the basic functions of Azure is to run applications. You're a developer, so naturally the "on what" of the execution is a fundamental one. Azure offers three models for running applications: virtual machines, websites, and cloud services.

Virtual Machines

The Windows Azure Virtual Machines service does pretty close to what you would expect it to do. That is, it enables you to create a virtual machine running in the cloud. Once the machine has been created, you can connect to the machine (using Remote Desktop Connection, for example) and manage the system as if it were sitting in the room next to you. This includes installing your own software, adding application roles, and applying patches and updates.

You can create an Azure virtual machine through a number of different mechanisms, including the Windows Azure management portal, a Windows PowerShell script, or a REST-based interface. You can select from a standard image (and the list of images includes Windows 2008 R2, Windows 2012, and a number of Linux distributions) or provide your own image.

What is almost nicer than the ease of creation is the payment model. As is true for most of the features in Azure, you pay by the hour. For the disk storage associated with the VM, you pay by the GB. In other words, you pay for what you use. This combination of functionality is, more or less, the Infrastructure as a Service that was mentioned earlier in the chapter.

Web Sites

If you look at the more common requests for cloud services, it shouldn't come as a surprise that hosting a website comes at the top of the list. Windows Azure Virtual Machines are certainly capable of hosting websites, but you are required to install the necessary application roles, manage security, and perform any other administrative functions that are required. Windows Azure Web Sites allow you to push the administrative tasks to someone else, specifically Microsoft.

A Windows Azure Web Site gives you a managed web environment that uses Internet Information Services (IIS). You can deploy most of your existing web applications (those that are currently running on IIS) into Azure Web Sites without making any changes. Or you can create a new website using one of the gallery images (which include WordPress, Joomla, and Drupal sites). You can also increase (or decrease) the number of instances of the website that are running so that spikes in traffic can easily be handled.

Cloud Services

The Cloud Services execution model is the one I think of as the traditional Azure model, because it is how you created a website right at the beginning of Azure. Originally (and still) the goal of cloud services was to provide an environment in which you could support many simultaneous users but, more importantly, could turn up or turn down the number of supported users very, very quickly. You can certainly achieve this goal with either Windows Azure Web Sites or Windows Azure Virtual Machines. However, for these other execution models, each of these requires more administrative effort than might be desired.

This is the space into which Cloud Services fits. Rather than writing a website in the traditional (at least for .NET) way, you segregate the functionality into Web roles and Worker roles. Web roles provide the Internet front-end. Worker roles provide background functionality. These different roles are then deployed onto a virtual machine. The difference between these VMs and the ones in the official Azure Virtual Machine service is the administrative effort. For the Cloud Service VMs, Azure takes care of all of the patching and operating system updating.

Each of these three execution models has its place in the architecting of a cloud-based system. In general, they fall along a continuum of administrative effort and level of control, ranging from the complete control of the Virtual Machine to the lack of control (at the operating system and machine level for Cloud Services). The execution models can be used separately or in conjunction with one another; the choice you make depends on the specific problem that you're trying to solve.

Data Storage

Whether you are using the cloud for hosting a web application or something entirely different, odds are very good that you will need to have access to some data. Windows Azure provides a number of different mechanisms to store and manage data, and, like the execution model, your choice depends on your needs.

Pick Your Own

Although it might be obvious, one of the first options you have is to use whatever database you want, so long as it can be hosted in an Azure Virtual Machine. Create a virtual machine that is running Window Server or Linux, install the database of your choice, and off you go. Naturally, you still have the administrative chores surrounding the operation of the VM. But if you require this level of flexibility, Window Azure is not going to get in your way.

SQL Database

Formerly known as SQL Azure, Windows Azure SQL Database provides the majority of functionality that you would expect from a SQL database. This includes transactional support, the ability to handle many concurrent users, and a common (and familiar) programming model. Though the analogy is not exactly 100 percent accurate, if you consider SQL Database to be SQL Server running in the cloud, you'd be pretty close. There is some functionality that is not available that would be part of an on-premises SQL Server environment. But for most scenarios (including the ability to use SQL Server Management Studio), SQL Database is sufficient.

Tables

In numerous cases, SQL databases are overkill for the situation. Sure, as developers we use SQL as our data store, but not because we need the full range of functionality. It's typically because that is the best available (or most familiar) choice. Table storage in Windows Azure provides another alternative for a specific set of scenarios.

The focus for Table storage is not on schemas, relationships, foreign keys and column constraints. Instead, Table storage is about the data and nothing but the data. The design focus is on being able to store large amounts of data and retrieve it quickly. This is not to say that relationships between tables are not possible. They are. But that is not the reason why Table storage is used.

Consider for a moment the situation in which you need to store user profile information. In general, this is not data that needs to be queried based on the values in the profile. For instance, it is not likely that an application using user profiles will need to get a list of the users that have the background color set to blue. More typically, the application will get the current profile data using only the user ID.

To implement this, you could create a SQL table, but then you need to define the columns that are included, along with the data type for each column. And if additional information is added to the profile, the SQL schema needs to be updated. This is a lot of effort when all you want to keep is a collection of properties for a particular user and retrieve or update it as needed.

This is the space where Windows Azure Tables are a good fit. They are basically key/value pairs where the value is a collection of properties (similar to a property bag, if you're familiar with the term). The data is accessed through the key only, allowing for fast retrieval, and a single table is capable of storing large quantities of data (upwards of a terabyte).

Blobs

For data where the items being stored tend to be large, Windows Azure Blobs offers a storage option similar to Azure Tables. In this case, the data being stored are large objects (think images, video, or backup files). They are presented to the application as if they were in a filesystem. However, they are actually in a key/value environment.

Connectivity

The holy grail of cloud computing might very well be to have your entire infrastructure hosted there. But that is, at best, a pipe dream. Not to mention that, even in the extreme cases, you would still need to have some machines not in the cloud, if only to use them to access the machines in the cloud. Fortunately (or naturally), Windows Azure offers a number of connectivity options to link your on-premises systems with those hosted in Azure.

Virtual Network

For many users, the servers that are running in their infrastructure are already virtual, at least from a mental perspective. Few users consider the physical location of the machines that host their SharePoint or SQL Server, or Exchange functionality. So long as the server is accessible to them through their network, they are happy to remain oblivious to its location. Nor should they ever really need to care.

Windows Azure Virtual Network provides this functionality to Windows Azure Virtual Machines. Using a VPN gateway device, system administrators can "extend" the internal network to the Virtual Machines. They can assign IP addresses to the VMs and configure them to be accessible locally with no additional user effort. The users will not need to know that they are accessing VMs in the cloud, because all of their desired functionality will "just work."

Azure Connect

For some scenarios, the requirement of configuring a VPN gateway device is too onerous. It might be that the requirement is to allow a Cloud Service application to access information stored on a single on-premises server. This is the space the Widows Azure Connect was intended to occupy. Instead of deploying a VPN gateway, a piece of software is installed on the on-premises machines. Once this has been done, the Azure applications can be configured to access those machines as if they were on the local network.

Traffic Manager

The Windows Azure Traffic Manager doesn't quite fit into the same functional niche as the features mentioned to this point. The purpose of Traffic Manager is not to link on-premises servers with cloud functionality. Instead, the goal of Traffic Manager is to help ensure that the users of your cloud functionality are connected to the most efficient location.

Windows Azure maintains a number of data centers all over the world, and as you deploy your cloud functionality you have the option as to which centers to use. Typically, you would like to have your users utilize the data center that is closest to where they are. However, if that most proximate center is overloaded, it would be useful to automatically route them to another data center. That is what Traffic Manager does for you.

Administrators define a set of routing rules related to distance, response time, and other factors. Traffic Manager is responsible for carrying out the rules to route incoming requests to the appropriate resource. If you are creating an application that scales globally, this is much desired functionality.

Authentication

Most applications want to know who you are. The biggest question that needs to be answered in that process is how much they trust your answer. Most applications want to check with another authority to determine the veracity of your response. For the Windows Azure world, the leading authority is Windows Azure Active Directory (AD).

The goal of Azure AD is pretty much the same as any other authentication provider. It takes the user's credentials, validates them, and returns a token that can be given to other applications. To help companies that have Active Directory running on-premises, Azure AD can synchronize account information between the corporate environment and the cloud.

Along with cloud-based AD services, Azure also provides Windows Azure Access Control Services. This tool can be used to help federate authentication with other popular identity providers, such as Facebook, Google, or Windows Account. Instead of forcing developers to learn the ins and outs of each of these different mechanisms, Access Control Services translates them into a single common format (OAuth) that forms the backbone of a single sign-on service.

Messaging

The need to interact with other applications is almost as common as the need for data. Although the number of options to do so are fairly numerous (direct calls and exposed service-based endpoints are two examples), Windows Azure is really only concerned with two message-passing patterns.

Azure Queues

As a concept, queuing is a simple one. In this incarnation, the queue is a first-in, first-out persistence mechanism. The idea is that one application places a message into a queue. That message is then read (at some future time) from the same queue, where it gets processed. In Windows Azure, this functionality is offered through Windows Azure Queues.

Though it is not exclusively an Azure concept (and that should not be a surprise to you), queues do form the backbone of the computational elements in Cloud Services. Queues are used to pass messages between the Web and Worker roles, which fits nicely into the scalability model that Cloud Services supports.

Service Bus

The Windows Azure Service Bus performs a function similar to Windows Azure Queues, yet with a number of subtle (and not-so-subtle) differences. Queues (not just Azure queues, but queues in general) operate as a communication between two processes. An individual message is read and processed by a single recipient. The Service Bus functions on a publish-and-subscribe model.

The sender posts a message to a previously created topic, and one or more subscribers to that topic receive a notification when the message is published. So, the communication is no longer one-to-one, but one-to-many.

In addition, Windows Azure Service Bus includes functionality such as order guarantees order (FIFO), delivery guarantees (at-most-once delivery), and transactions. Beyond these distinctions, requirements such as performance, capacity and security can play a role in determining whether Azure Service Bus or Azure Queues is the most appropriate choice.

AND NOW, BACK TO OUR SHOW

To this point, I have barely mentioned Windows Azure Mobile Services (WAMS). It seems odd to be most of the way through the first chapter and the topic has not yet been described. Let's take a few moments to rectify that situation.

Windows Azure Mobile Services is an interesting combination of functionality. At its basic level, it is a prebuilt, preconfigured website and database combination that is used solely to perform CRUD operations on data. Beyond that, it has a number of different twists, as well as some additional features. But when you look at the underpinnings, you see a service that can be used to store and retrieve data through an interface that is readily available across all mobile platforms. The strength of WAMS lies in this basic function (one that is needed by almost all non-trivial mobile apps), the stability and scalability offered by Azure, and some of the "extras" that come along for the ride.

Setting Up WAMS

Start by getting WAMS set up so that it's ready for use. The starting point is a Windows Azure account. Using your favorite browser, navigate to http://www.windowsazure.com. Once you have logged in with a Microsoft Account (formerly a Microsoft Live ID); you should see the Windows Azure homepage. Clicking the Portal link (in the top right) takes you to the Windows Azure Management Portal (see Figure 1-1).

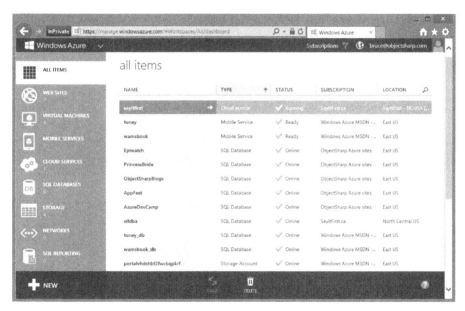

FIGURE 1-1

As mentioned, to access Windows Azure, you do need to have a Microsoft Account, which means that you need to create one if you don't already have one. Go to https://account.live.com and click the Sign Up Now link at the bottom right of the screen.

Once you are in the portal, you see a collection of icons on the left side of the screen. Each icon represents a feature or function within Azure. To create your mobile service, look for the Mobile Services icon and click it.

At the time of writing, Windows Azure Mobile Services was still in preview mode. So if you can't find the Mobile Services icon, you need to activate the feature first. Navigate to `https://account.windowsazure.com/PreviewFeatures`, locate the Mobile Services section and click on the Activate link. Once the feature has been activated (and it may take a few minutes), the Mobile Services icon will be visible in your portal.

Once you have clicked on the Mobile Services icon the screen that appears is used to manage all of your mobile services. At the bottom left of the screen you see a New icon. This icon is used to create any Azure artifact. Click it to display the new item pane (Figure 1-2).

FIGURE 1-2

The new item pane is constructed using a horizontal menu structure. As you select an item on the left the option immediately to the right changes. Select an item from the second set of options, and either another collection of choices or a dialog box appears. To create a mobile service, start by selecting the Compute option. Then choose Mobile Service from the choices that appear to the right. Finally, select the Create option from the choices that appear to the right of Mobile Services. The New Mobile Service dialog box appears (Figure 1-3).

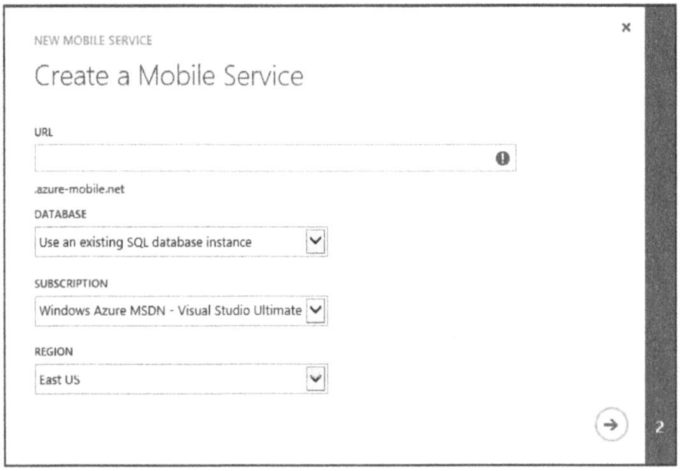

FIGURE 1-3

The first page of the New Mobile Service dialog box lets you specify some fundamental information about your mobile service. For instance, your mobile service must be given a name. This name has some limitations, because of how it will be used to identify your mobile service. Specifically, it will appear as the header portion of the URL that uniquely identifies your service. So if you selected a name of WAMSBook, the endpoint for your service would be http://wamsbook.azure-mobile.net.

The result of using the name of the service in the URL means that the name has to adhere to some naming standards. It must be between 2 and 60 characters in length. It must start with a letter and contain only letters, numbers, and dashes. And, as a potential roadblock, it must be unique. That is, unique across all mobile services created by anyone.

Along with the name of the mobile service, you must provide three other values. The first is the database. WAMS has a database that enables you to persist information for use within the service. For the Database drop-down, you have the choice of creating a new database or using an existing database instance. The decision here has to do with whether or not you have an active Azure SQL Database service. If you do, you can use that component to persist your data. Otherwise, a new instance will be created for you.

The Subscription drop-down contains a list of the Azure subscriptions for which you are an administrator. The selection you make determines who is responsible for paying for any charges related to the service. As of this writing, it is possible to create and use mobile services at no cost. However, limits exist on the number of resources that are used by the free version. If it turns out that those limits are not sufficient for your needs, you can upgrade to a paid-for version. In that case, the selected subscription is the one that is charged.

The final piece of information is the region into which the mobile service is deployed. Here, your choices are the list of regions that support mobile services. As a general rule, you would select the region that is geographically closest to where the consumers of your application are found.

Once you have made selections for all of these values, select the arrow at the bottom right of the dialog box. This causes the second page of the dialog box to appear (Figure 1-4).

The details of what you see on this second page depend on one of the choices you made on the first page — specifically, the database that you want to use. If you indicated that you wanted to use an existing database instance, you see the dialog box on the top of Figure 1-4, where you need to choose the database and provide the necessary login credentials. If you chose to create a new database instance, you see the dialog box on the bottom of Figure 1-4, where you need to provide the name for the new database instance, and choose the server on which the database will be placed. If you select an existing server, you also need to provide valid credentials to the server. For a new server, you can specify the login name and password, as well as the region in which the server will be placed. Once you have made the necessary choices and provided all of the information that Azure needs, select the check button in the bottom right of the screen to create your mobile service.

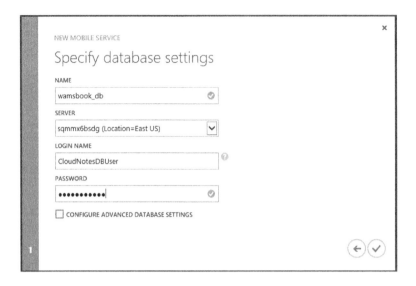

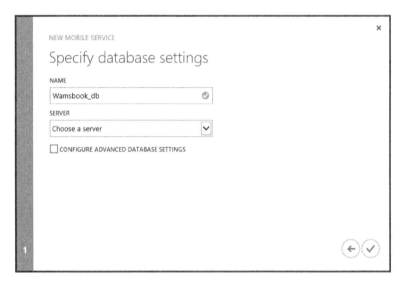

FIGURE 1-4

After a few minutes (and probably less time than you might expect), your mobile service is active and ready to go. You will see it appear in the list of mobile services as illustrated in Figure 1-5.

And Now, Back to our Show

FIGURE 1-5

Generating a Sample Application

You have just finished setting up and configuring your mobile service. Figure 1-5 shows the new mobile service in the list. Although you don't have to go through these next steps, it is useful if this is your first exposure to Mobile Services. Click your newly minted mobile service and you are taken to the Quick Start view shown in Figure 1-6.

FIGURE 1-6

The Quick Start view is a nice place for people who are unfamiliar with Mobile Services to get started. At the bottom, you see a couple of links that take you to MSDN Library documents describing how to do some common tasks (tasks that are covered later in this book in Chapters 4 and 6, specifically). You also see two links that are of interest as you create a simple sample application for using Mobile Services.

Samples are available for Windows Store, Windows Phone 8, and iOS applications. In the Choose Platform section, select the platform that you would like to generate a sample for. The instructions in the next few paragraphs presume that you have selected Windows Store as the platform, but the basic steps are the same for each platform.

After selecting Windows Store as the platform, click the Create a New Windows Store App link. This is a three-step process for creating the app. First you need to get the appropriate tools. As a reader of this book, there is a small assumption that you already have a version of Visual Studio 2012 available to you. However, if you don't, you can download the free Express version of Visual Studio 2012 at http://www.microsoft.com/visualstudio/eng/downloads#d-express-windows-8. Even if you already have Visual Studio installed, you still need to get the Mobile Services SDK, which you can download at http://www.windowsazure.com/en-us/develop/mobile/developer-tools/. Make sure you do so before running the sample applications.

The second step creates a table in your mobile service. More precisely, there is a button that, when clicked, creates a TodoItem table in the mobile service database. As you see in Chapter 2, the data that is stored in your mobile service does get stored in a table, and coding elements are used to map classes onto the columns in the table. But underlying the data persistence function of WAMS is a SQL table.

The third step is to download the sample application. You can choose whether the language for the sample application is C# or JavaScript. Then, when you click the Download button, a ZIP file (politely named for your mobile service) is downloaded. Opening the ZIP file reveals a Visual Studio solution and related files for your sample application. Unpack the contents of the ZIP file and open the solution file. If you have selected a Windows Store application, you may be prompted to get a Developer License. This is required to develop Windows Store applications, and involves agreeing to some terms and providing your Microsoft Account credentials.

Run the application (no need to make any changes) and you will shortly see the application shown in Figure 1-7.

No question that the application is a simple one. However, it does enable you to create a TodoItem, which is then persisted to the TodoItem table in your mobile service. It also provides a mechanism that marks a particular TodoItem as being complete. After you have created and completed a number of TodoItems, go back to the Mobile Services page in your browser. Click the Data tab and a list of the tables defined in your mobile service is shown. For the sample application, that is the TodoItem table created way back in the second step of this process. When you click the TodoItem table, the details for that table are displayed (Figure 1-8).

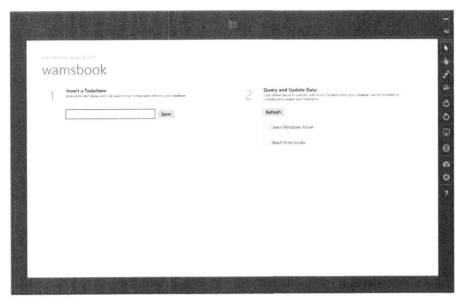

FIGURE 1-7

FIGURE 1-8

Although a couple of tabs are available for each table, the default view is to show the records. You will see the TodoItems that you created through your sample application appear in the list.

SUMMARY

By now, you have not only learned about the different components that make up the Windows Azure platform, but also created your first mobile service, along with a client application that takes advantage of it. Over the next few chapters, you take a look at the functionality that Windows Azure Mobile Services exposes on both the client and server side. This functionality enables you to rapidly develop and deploy a number of fairly complex (and useful) scenarios.

2
Creating and Manipulating Data

IN THIS CHAPTER:

- ➤ Understand how the WAMS data model can be created using dynamic schema generation
- ➤ Learn the methods used to perform CRUD operations on a mobile service table
- ➤ Examine the sample application to see how the methods are put into action

WROX.COM CODE DOWNLOADS FOR THIS CHAPTER

The wrox.com code downloads for this chapter are found at http://www.wrox.com/go/windowsazuremobileservices on the Download Code tab. The code is in the Chapter 2 download, and individually named according to the names throughout the chapter.

Now you have a mobile service sitting out there and ready for you to use. Although a number of other features are covered as you go deeper into the book, the fundamental function of Windows Azure Mobile Services (WAMS) is to provide a mechanism for storing and retrieving data from a mobile platform. Yes, tools are available that allow push notifications to be sent out when data is modified. Yes, you can create server-side scripting to extend or validate incoming requests. Yes, it's even possible to run tasks in Mobile Services on a scheduled basis. However, the starting point for any soup-to-nuts examination of WAMS needs to be how to create, update, and retrieve data.

The manipulation of data using WAMS is a two-step process. The first involves the creation of the data model. To be fair, this isn't an absolute requirement. In Chapter 5 – Using REST to Access WAMS Data, you see how to retrieve data using only HTTP requests, and you can do that without a preexisting data transfer class. However, the Windows Azure Mobile Services SDK includes a client-side proxy that you can use to access the stored data. Through that proxy, an attribute-based mechanism enables you to map .NET classes onto tables in your mobile service.

THE DATA MODEL

The attributes that define the data model used by the WAMS proxy class are straightforward. As a starting point, consider the following class definition:

C#

```
[DataTable]
public class GameScore {
   [DataMember]
   public int Id { get; set; }

   [DataMember]
   public int Score { get; set; }

   [DataMember]
   public DateTime GameDate { get; set; }
}
```

VB.NET

```
<DataTable> _
Public Class GameScore
   <DataMember> _
   Public Property Id As Integer

   <DataMember> _
   Public Property Score As Integer

   <DataMember> _
   Public Property GameDate As DateTime
End Class
```

When this class is used to access WAMS, the data is inserted into, updated in, deleted from, or retrieved from a table called GameScore. In addition, the columns in the WAMS table are named Id, Score, and GameDate.

One of the interesting aspects of WAMS, at least in terms of being able to get up and running quickly, is that you can configure your mobile service to dynamically create the underlying data store as needed. That is, the first time you insert data into a brand new mobile service table, WAMS creates the table referenced by the name of your class. It then creates the columns represented by the properties in your class. If, at some point in the future, you add a property to the class, WAMS adds a column into the data store to represent the new property.

To give you a bit of a peek under the covers, take a look at what WAMS does with this dynamic creation of columns. When WAMS sends your object across the wire from the client to the mobile service, it gets converted into JSON. When the JSON object is received, the mobile service examines the properties to see if the column already exists in the data store. If it doesn't, WAMS creates a column. Although WAMS could just create a text column for each property (because, after all, the properties in JSON objects are transmitted as strings), it does something a little more appropriate. Table 2-1 illustrates the mapping between JSON data types and SQL column types.

TABLE 2-1: JSON to WAMS SQL Azure Data Types Mapping

JSON / CLR DATA TYPE	T-SQL TYPE IN UNDERLYING SQL AZURE DATABASE TABLE
Numeric values (integer, decimal, floating point)	float(53) – The highest precision data type
Boolean	Bit
DateTime	DateTimeOffset(3)
String	nvarchar(max)

To avoid confusion when you look at the console for your mobile service, the data columns are displayed as one of number, date, string, or Boolean to reduce complexity. The conversion between these types and the underlying T-SQL type is handled automatically by WAMS.

Dynamic schema generation is quite a useful option. At least, it is to a point. Once you deploy your application into production, it seems unlikely that your schema would or *should* change. As a result, you can turn this capability on and off. It is recommended that for a production environment you *don't* leave this capability turned on. To toggle this functionality on or off, go to the Windows Azure portal and get to the dashboard for your mobile service. Click the Configure menu item to reveal the screen shown in Figure 2-1.

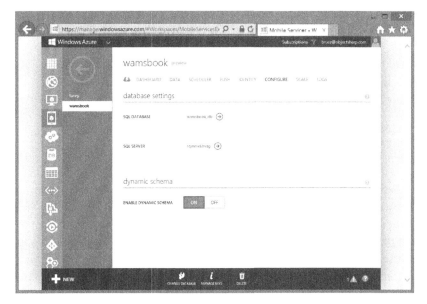

FIGURE 2-1

At the bottom of the screen, you see an option titled Enable Dynamic Schema. You can select On or Off as you need, making sure to save your changes when finished.

When you are dynamically creating the schema for the data store, the name of the table and columns come from the class that is used as the *Data Transfer Object* (DTO). However, sometimes that might not be appropriate. Some scenarios that fall into this category are when you are using an existing database and the column or table names don't match the names you want to use for your class. Or, if the class that defines the DTO already exists and you want the names in your mobile service to be slightly different.

Fortunately, the DataTable and DataMember attributes enable you to customize the table and property names used in your mobile service. Consider the following class definitions, which show the previous class with some minor changes:

C#

```
[DataTable(Name="gameResult")]
public class GameScore {
    [DataMember(Name="id")]
    public int Id { get; set; }

    [DataMember(Name="score")]
    public int Score { get; set; }

    [DataMember(Name="resultDate")]
    public DateTime GameDate { get; set; }
}
```

VB.NET

```
<DataTable(Name="gameResult")> _
Public Class GameScore
    <DataMember(Name="id")> _
    Public Property Id As Integer

    <DataMember(Name="score")> _
    Public Property Score As Integer

    <DataMember(Name="resultDate")> _
    Public Property GameDate As DateTime
End Class
```

The difference between this class declaration and the one that appeared earlier in the section is the use of the Name property. This property is used in the DataTable and DataMember attributes to change the name of the table or properties used on the server-side of WAMS. In this particular case, a table named gameResult is created with columns of id, score, and resultDate. Note that the DataTable attribute is found in the Microsoft.WindowsAzure .MobileServices namespace. However, the DataMember attribute, which is the same one that is used in other serialization methods, is found in the System.Runtime.Serialization namespace.

The use of the `DataTable` and `DataMember` attributes are not required to create a data transfer class. They have been explicitly included in the code snippets for clarity. However, if your data transfer class has a property that should not be included in your mobile service, it needs to be marked with the `IgnoreDataMember` attribute.

C#

```
[IgnoreDataMember]
public string UnpersistedData{ get; set; }
```

VB.NET

```
<IgnoreDataMember> _
Public Property UnpersistedData As String
```

I need to mention one other aspect of the DTO. Keep in mind that the basic functionality of WAMS is the storage and retrieval of data. The storing of the data isn't a big deal, but when it comes to retrieval there is an obvious (although not always considered) question of how to uniquely identify the items that are to be retrieved. In the world of database design, you typically have two ways to automatically generate the key for a record: GUIDs and identity columns. Although theoretically, either one of these could work, the people who created WAMS decided to use identity columns. The identity column in the preceding classes is the `Id` property. It is a requirement that any class used in WAMS includes an `Id` property with a type of integer that is used to uniquely identify the record.

If you don't want to create the schema dynamically (the WAMS version of model-first development); you can create the tables and columns through the Azure Management Portal. You can also view the data that is currently stored in the table.

To start, go to the Azure Management Portal (after you have logged in, naturally), click the Mobile Services icon, and select your mobile service from the list that appears. This takes you to your service's dashboard or Quick Start screen. Click the Data tab to view the tables that are part of your mobile service (Figure 2-2).

You can modify the attributes associated with a given table by clicking the table in the list, or you can create a new table (which is what you're about to do) by clicking the Create link. When you click the link, the Create New Table dialog box appears (Figure 2-3).

The main attribute of the table is the name of the table itself, and the table name has the same constraints as any SQL Server table. You need to map the table map onto the class name or the name property in the `DataTable` attribute for the class. For this example, use the name of `gameResult`, so that it maps nicely onto the class that appears earlier in the chapter.

Along with the table name, you can set four permissions. These permissions control who has access to the Insert, Update, Retrieve, and Delete functions for the table. You identify the permission on each of these functions by setting the value associated with each function to one of the following choices:

FIGURE 2-2

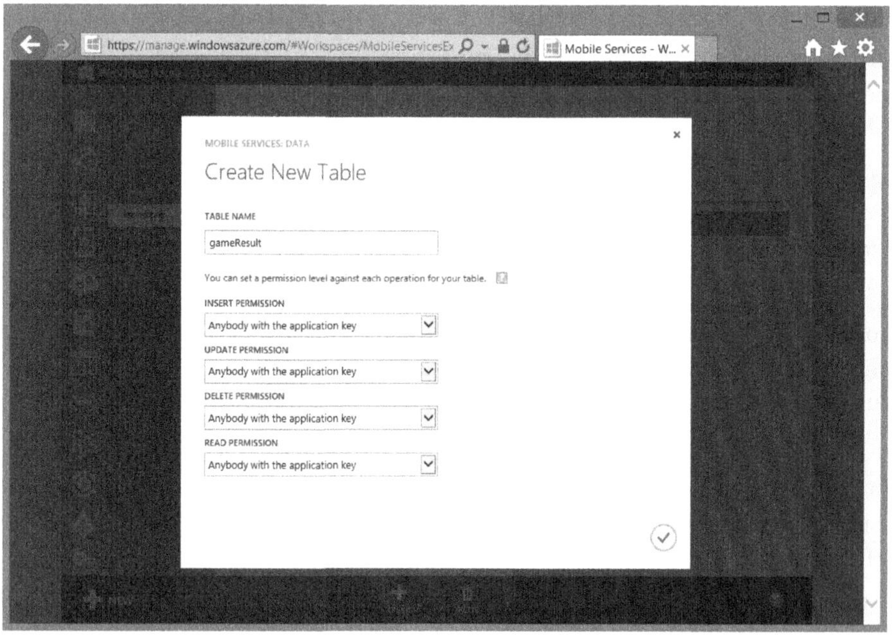

FIGURE 2-3

- **Everyone** — This is probably the most straightforward of the choices. By setting the permission to Everyone, you are allowing every user to perform the function on the table.
- **Anybody with the Application Key** — I talk about the application key in a moment, but functionally this means that any request that includes the application key is able to perform the function on the table. In practice, this means "your application" is able to perform the function. It is also the default permission whether the table is created dynamically or through the Management Portal.
- **Only Authenticated Users** — If the user making the request has been authenticated, he or she is allowed to perform the function on the table. The process of authenticating users is covered in Chapter 4.
- **Only Scripts and Admins** — The function on the table can only be accessed by scripts (covered in Chapter 3) or by administrators. Administrators are identified by the inclusion of a special master key in the request.

Once you click the check icon in the bottom right of the form, the table is created. However, the table itself has only a single column (the Id). Adding more columns is something that is of immediate importance and practicality.

You can add additional columns in three ways. First, you can turn the dynamic schema option on (as described earlier in this chapter), create a class with a name that matches the table name, properties that match the columns that you want to have, and insert a record. How to insert the record hasn't been covered yet (patience…it's later in this chapter), but the insertion of the first record creates the defined columns.

If that seems a little too "fly by the seat of your pants" for you, there is a second alternative. Underlying your mobile service is a SQL database. You can use SQL Server Management Studio to connect to the database. Once connected, you can modify the data or the data schema as you see fit.

The third mechanism is to use a tool available through the Azure Management Portal to modify the database. In the main Azure Management Portal page, click SQL Databases on the left. This displays a list of the SQL databases that are part of your Azure account (Figure 2-4).

Included in this list of databases is the one that was created when you initially set up your mobile service. Select that database and click the Manage button that is visible at the bottom of the screen.

If this is the first time you are managing your database from this IP address, you are prompted to add the IP address to the list of those that are allowed to manage the database. One of the security mechanisms that is implemented with Azure SQL databases is port blocking. To be able to access the database from outside of Azure, you need to add a rule that allows communications to pass through the firewall from a specific IP address. This is true even if you are using SQL Server Management Studio to modify the database, so consider yourself warned.

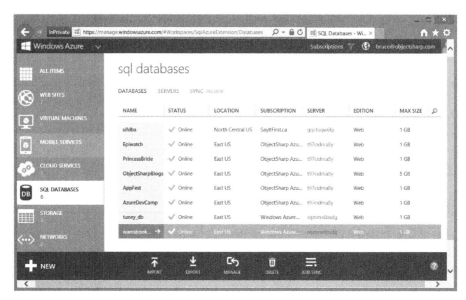

FIGURE 2-4

When you have added your IP address, a prompt confirming that you want to manage the database appears. Click Yes and then log in to the database management screen using the credentials that you provided when you set up the mobile service. The result is a screen that looks like Figure 2-5.

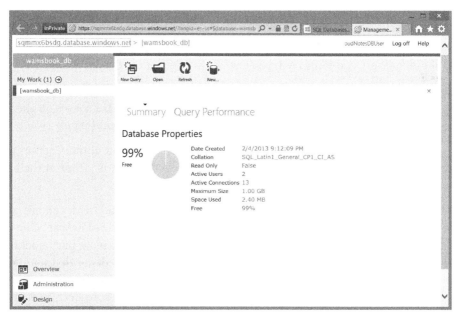

FIGURE 2-5

The Data Model

On the left side, you see a number of options. For what you're trying to do, click the Design option. A list of the tables in the database appears (Figure 2-6).

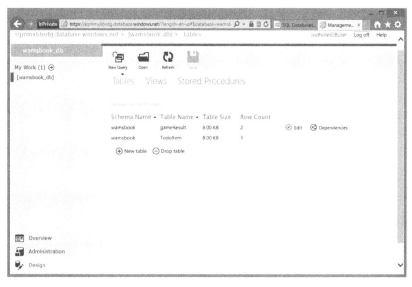

FIGURE 2-6

Next to each table is an Edit icon which, when clicked, enables you add or modify the columns in the table. Click the Edit icon next to the table in your mobile service that you want to modify as the screen shown in Figure 2-7 displays.

FIGURE 2-7

In this screen, you can remove a column by selecting the column to delete and clicking the Delete Column icon. If you want to add a column, click the Add Column icon, then change the column's name, type, default value, and whether it's an identity value, the primary key, or is required. When you have finished making the desired changes, click the Save icon at the top of the screen.

A couple of limitations exist when it comes to modifying the database that is used by your mobile service. First, you should *not* modify the Id column. It is a critical part of how WAMS does what it does, and attempting to change it is asking for trouble (if not immediately, then down the road). The second condition is that you should be careful when you change the database. Technically speaking, although accessing the database is supported, support is not provided if you have changed the schema and your mobile service stops working. So it is technically possible to do something that causes WAMS to stop functioning (at least with this table). You are free to add columns, indices, or even triggers to the tables that support WAMS. Just do it carefully.

Client-Side Functionality

Now that you have seen the various ways that you can create the data model for your mobile service, the time has come to actually put it into practice. The following sections look at the techniques you can use to perform CRUD functions on the data. In each case, I demonstrate the coding used in a number of different languages and environments. Specifically, I cover the .NET stalwarts of C# and VB.NET, as well as JavaScript and Objective C, which are relative newcomers (at least to WAMS).

Getting the Application Key

The starting point for communicating with your mobile service is to get the application key. Though it is possible to configure your tables so that the application key is not required (see the section on setting permissions earlier in this chapter), most of the time an application key is necessary.

If you download the sample application as described in Chapter 1 and look through the code, you will find that the application key for your mobile service is already present. Specifically, it is in the App.xaml code file where the MobileService field is declared. However, this section describes how to find the application key through the Azure Management Portal.

From within the Azure Management Portal, click the Mobile Services icon on the left to display the list of your mobile services. Then click the desired mobile service to display either the dashboard or the Quick Start screen for that service. If the Quick Start screen is displayed, click the Dashboard tab at the top to get to the screen shown in Figure 2-8.

One of the pieces of information you need to communicate with your mobile service is the Site URL on the right side of the screen (indicated with the arrow in Figure 2-8). Keep note of the value for use in the next section. To continue the process of retrieving your application key, click the Manage Keys icon at the bottom of the Figure 2-8.

FIGURE 2-8

The display that appears (shown in Figure 2-9) contains the magical application key that has been the object of your quest. It also contains the master key, which is used to indicate that the mobile service is being accessed by someone with elevated permissions (conceptually, an administrator). As with the Site URL, make note of the application key for use in the next section.

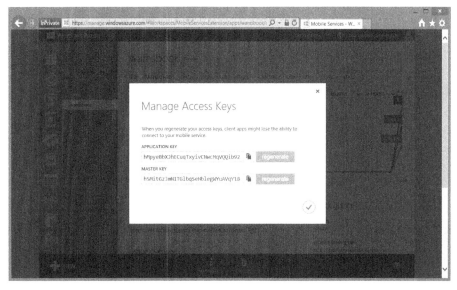

FIGURE 2-9

Constructing the Proxy Object

Now that you have an application key, it's time to put it to use. For the .NET code that is used in the rest of this chapter (and, in most cases, throughout the rest of the book), you need to install the Mobile Services SDK (which you can download at http://www.windowsazure.com/en-us/downloads. This SDK contains the declarations and implementation for a number of different classes whose purpose is to reduce the amount of code you write.

If you create a new project that wants to use WAMS, you need to add a reference an assembly that was installed with the Windows Azure Mobile Services SDK. This assembly is named Windows Azure Mobile Services Managed Client. However, the following code snippets are part of the sample application, which was downloaded in the "Generating a Sample Application" section of Chapter 1 – Introduction and Fundamental Concepts.

In the App.xaml code file there is a class named App, which is intended to be (among other things) a container for values that are available across the application. For your application, create a publicly exposed, read-only field named MobileService. It is through this field that access to the data tables in your mobile service will take place. The following code illustrates the creation of this field:

C#
```
public static MobileServiceClient MobileService = new MobileServiceClient(
            "https://wamsbook.azure-mobile.net/",
            "FGNIFOZqbYRpLIHVlbzCMsCthJxala84"
            );
```

VB.NET
```
Public Shared MobileService As MobileServiceClient = _
   New MobileServiceClient( _
      "https://wamsbook.azure-mobile.net/", _
      "FGNIFOZqbYRpLIHVlbzCMsCthJxala84" _
   )
```

A quick look at the second parameter of the constructor reveals a collection of what appears to be gibberish. This gibberish should be the application key you found in the previous section. To enable your application to communicate with your mobile services, you need to replace the second parameter with the key you found earlier.

Inserting a Record

The basic flow for performing insertions, deletions, or modifications in WAMS is the same:

- ▶ Build (or retrieve) the target object
- ▶ Get a reference to the data table
- ▶ Invoke the appropriate method asynchronously, passing in the object

For the creation of a new record, you invoke the `InsertAsync` method. Consider the following code:

C#
```
GameScore gameResult = new GameScore
    {
        Score = 145,
        GameDate = DateTime.Now
    };
await App.MobileService.GetTable<GameScore>().InsertAsync(gameResult);
```

VB.NET
```
Dim gameResult as New GameScore()
gameResult.Score = 145
gameResult.GameDate = DateTime.Now
App.MobileService.GetTable(Of GameScore)().InsertAsync(gameResult)
```

The flow of the code was described earlier: the object is created, a reference to the data table is created (this is the result of the `GetTable` generic method), and the `InsertAsync` method is invoked.

Relatively speaking, this is a very easy and intuitive way to add a record to your mobile service, but two points are important to notice. First, the type provided as the generic portion of the `GetTable` call is used to determine which data table is to be updated. The name of the class is the source of this information. If the class has been decorated with the `DataTable` attribute and the `Name` property used, then the non-generic version of `GetTable` can be used. That version takes the name of the table as the parameter. So the following two statements will return a reference to the same table, based on the class declaration found earlier in this chapter:

C#
```
var gameScoreTable = App.MobileService.GetTable<GameScore>();
var gameResultTable = App.MobileService.GetTable("gameResult");
```

VB.NET
```
Dim gameScoreTable = App.MobileService.GetTable(Of GameScore)()
Dim gameResultTable = App.MobileService.GetTable("gameResult")
```

The second point relates to the `Id` property on the `GameScore` object. Notice that it is not assigned a value. This is because, as a new record, there is no previously assigned `Id`. Because it is assigned automatically by the mobile service (if you recall, the `Id` property is marked as an Identity column), there is no reason to pass it along. If you were to check on the value of the `Id` property in the object after the call returns, you would find that it contains the value of the `Id` for the just-added record.

Deleting a Record

Deleting a record is even simpler than inserting one. Consider the following code:

C#

```csharp
GameScore gameResult = new GameScore
    {
        Id = 213
    };
await App.MobileService.GetTable<GameScore>().DeleteAsync(gameResult);
```

VB.NET

```vbnet
Dim gameResult as New GameScore()
gameResult.Id = 213
App.MobileService.GetTable(Of GameScore)().DeleteAsync(gameResult)
```

The difference between insertion and deletion is that the record being passed to the `DeleteAsync` method has an `Id` assigned. Notice also that the rest of values are not filled in. They can be, but it is not a requirement of the default behavior of the method.

Modifying an Existing Record

You might be seeing a pattern here, and it continues with the modification process. Consider the following code:

C#

```csharp
GameScore gameResult = new GameScore
    {
        Id = 213,
        Score = 145,
        GameDate = DateTime.Now
    };
await App.MobileService.GetTable<GameScore>().UpdateAsync(gameResult);
```

VB.NET

```vbnet
Dim gameResult as New GameScore()
gameResult.Id = 213
gameResult.Score = 145
gameResult.GameDate = DateTime.Now
App.MobileService.GetTable(Of GameScore)().UpdateAsync(gameResult)
```

This is very similar to the previous examples in terms of structure. It's also not particularly realistic when you consider how you would typically update a record. The more typical approach would be to retrieve an existing record from the mobile service, make the desired changes to the property values, and then call `UpdateAsync`. However, the code does illustrate the basic flow, and it illustrates that, ultimately, the `Id` is the mechanism through which existing records are identified with it comes to making changes.

Retrieving Data

The retrieval of data from your mobile service veers away from the pattern you have seen to this point. The starting point is still the same. You get a reference to the WAMS data table using the `GetTable` generic method. This object (which is actually an implementation of the `IMobileServiceTable` generic interface) exposes a `Where` method. The `Where` method takes a `Predicate` object as a parameter, and the `Predicate` object is an expression that evaluates to a Boolean value. Records in the mobile service for which the expression evaluates to `true` are included in the list of resulting objects. Here's an example to illustrate:

C#

```
var table = App.MobileService.GetTable<GameScore>();
var resultList = table.Where(g => g.Score > 500);
foreach (GameScore result in resultList) {
   // do something with each GameScore object
)
```

VB.NET

```
Dim table = App.MobileService.GetTable(Of GameScore)()
Dim resultList = table.Where(Function(g) g.Score > 500)
Dim result as GameScore
For Each result In resultList{
   // do something with each GameScore object
Next For
```

If you are comfortable with LINQ, this code will look quite familiar. When the `Where` method is executed, it returns an `IEnumerable` of `GameScore` objects that have a game score greater than 500. To be precise, as a result of the delayed execution model of LINQ, the retrieval of the data won't occur until the `For Each` loop is processed.

Now, there is a problem with this example that makes it less than optimal for production applications. When the data is retrieved, it is done synchronously. If this code runs on the UI thread, your application will appear to freeze, which is not the best user experience.

You can easily correct this problem, however. The following code snippet implements the same functionality, but in an asynchronous manner:

C#

```
var table = App.MobileService.GetTable<GameScore>();
var resultList = await table.Where(g => g.Score > 500).ToListAsync();
foreach (GameScore result in resultList) {
   // do something with each GameScore object
)
```

VB.NET
```
Dim table = App.MobileService.GetTable(Of GameScore)()
Dim resultList = table.Where(Function(g) g.Score > 500).ToListAsync()
Dim result as GameScore
For Each result In resultList
    // do something with each GameScore object
Next For
```

An Example

Now that you have the basics, take a look at an example that uses these methods. Start by creating a project in Visual Studio. I presume that you're working with a Windows Store project, but you could run the same code using a Windows Phone application as well.

Also, as you add code to your project, some of the types might appear with red, squiggly underlines. This is because you haven't added the appropriate using or Imports statement to the code file. Rather than describe all of these additions, the presumption is that you are able to add them as needed.

Before starting on the code, add a reference to your project to the Mobile Services Client assembly. You do this by right-clicking the project in Solution Explorer and selecting Add Reference. In the Reference Manager dialog box, navigate to the Windows ➪ Extensions node (on the right) and double-click the Windows Azure Mobile Services Managed Client that appears in the list. Click OK at the bottom of the dialog box to complete the addition.

Open the MainPage.xaml file from within the Solution Explorer and add the following XAML to the Grid element that is part of the template. This places a text box onto the form, along with a couple of buttons and a list box to display information.

```
<Grid.RowDefinitions>
    <RowDefinition Height="50" />
    <RowDefinition Height="50" />
    <RowDefinition Height="50" />
    <RowDefinition Height="*" />
</Grid.RowDefinitions>
<Grid.ColumnDefinitions>
    <ColumnDefinition Width="*" />
    <ColumnDefinition Width="2*" />
    <ColumnDefinition Width="3*" />
    <ColumnDefinition Width="*" />
</Grid.ColumnDefinitions>
<TextBlock Grid.Row="1" Grid.Column="1" Text="Score:" FontSize="32"
    TextAlignment="Right" Margin="5" />
<TextBox Grid.Row="1" Grid.Column="2" Name="tbxScore" Margin="5"
    HorizontalAlignment="Stretch" VerticalAlignment="Stretch" />
<StackPanel Grid.Row="2" Grid.Column="1" Grid.ColumnSpan="2"
    Orientation="Horizontal" HorizontalAlignment="Center">
    <Button Content="Add Score" Margin="5" FontSize="18" Name="btnAddScore" />
```

```xml
    <Button Content="Update Score" Margin="5" FontSize="18" Name="btnUpdateScore" />
    <Button Content="Delete Score" Margin="5" FontSize="18" Name="btnDeleteScore" />
</StackPanel>
<ListBox Grid.Row="3" Grid.Column="2" Name="lbxGameScores" Margin="5">
    <ListBox.ItemTemplate>
        <DataTemplate>
            <StackPanel>
                <TextBlock Text="{Binding GameDate}" />
                <TextBlock Text="{Binding Score}" />
            </StackPanel>
        </DataTemplate>
    </ListBox.ItemTemplate>
</ListBox>
```

Finally, as part of the setup, add a GameScore class declaration to your project. Right-click the project in Solution Explorer and select Add ➪ Class. In the Add Class dialog box, set the name to **GameScore** and click the Add button. In the GameScore file that is created, change the declaration of the GameScore class to be the following (this is a duplicate of code that appeared earlier in the chapter, but is included here for convenience):

C#

```csharp
[DataTable(Name="gameResult")]
public class GameScore {
   [DataMember(Name="id")]
   public int Id { get; set; }

   [DataMember(Name="score")]
   public int Score { get; set; }

   [DataMember(Name="resultDate")]
   public DateTime GameDate { get; set; }
}
```

VB.NET

```vbnet
<DataTable(Name="gameResult")> _
Public Class GameScore
    <DataMember(Name="id")> _
    Public Property Id As Integer

    <DataMember(Name="score")> _
    Public Property Score As Integer

    <DataMember(Name="resultDate")> _
    Public Property GameDate As DateTime
End Class
```

Now wire up the mobile service. Open the code for the App.xaml file. In it, you'll see a definition for the App class. You're going to add a static field to this class that references your mobile service. Add the following code at the top of the App class:

C#
```
public static MobileServiceClient MobileService = new MobileServiceClient(
            "https://wamsbook.azure-mobile.net/",
            "FGNIFOZqbYRpLIHVlbzCMsCthJxala84"
            );
```

VB.NET
```
Public Shared MobileService As MobileServiceClient = _
   New MobileServiceClient( _
      "https://wamsbook.azure-mobile.net/", _
      "FGNIFOZqbYRpLIHVlbzCMsCthJxala84" _
   )
```

Of course, the second parameter to the constructor needs to be the value of the application key that you identified in the "Getting Your Application Key" section earlier in this chapter. And the first parameter needs to be the URL to your mobile service.

You need to add some `Click` event handlers to the buttons that appear in the UI. In the designer for the MainPage.xaml file (the one where you added the XAML a few moments ago), double-click the Add Score button. This creates the event handler for the `Click` event.

Inside this event, add the code to create a new `GameResult` class and pass it to the appropriate `InsertAsync` method. Also, since the call to the `InsertAsync` method uses the `await` keyword, the event handler method needs to have the `async` keyword added to the declaration. The event handler should look like the following:

C#
```
private async void btnAddScore_Click(object sender, RoutedEventArgs e)
{
   currentScore = new GameScore()
   {
      GameDate = DateTime.Now,
      Score = Convert.ToInt32(tbxScore.Text)
   };
   await App.MobileService.GetTable<GameScore>().InsertAsync(currentScore);
   refreshList();
}
```

VB.NET
```
Private Async Sub btnAddScore_Click(sender As Object, e As RoutedEventArgs)
   currentScore = New GameScore()
   currentScore.GameDate = DateTime.Now
```

```
        currentScore.Score = Convert.ToInt32(tbxScore.Text)

        Await App.MobileService.GetTable(Of GameScore)().InsertAsync(currentScore)
        refreshList()
End Sub
```

The event handler method assigns a new instance of the GameScore class to a variable named currentScore. You need to define this variable at the class level. To do this, add the following code at the top of the MainPage class:

C#
```
GameScore currentScore;
```
VB.NET
```
Dim currentScore As GameScore
```

You'll also notice that there is a call to the refreshList method at the end of the Click event handler. This method, which you are about to create, retrieves all of the GameScore objects currently in your mobile service. Once retrieved, the objects are bound to a list box for display purposes. Add the following method to the MainPage class:

C#
```
private async void refreshList()
{
    List<GameScore> scores = await
        App.MobileService.GetTable<GameScore>().Select(gs => gs).ToListAsync();
    lbxGameScores.ItemsSource = scores;
}
```
VB.NET
```
Private Async Sub refreshList()

    Dim scores As List(Of GameScore) scores = await _
       App.MobileService.GetTable(Of GameScore)() _
       .Select(Function(gs) gs).ToListAsync()
    lbxGameScores.ItemsSource = scores

End Sub
```

While you're at it, add the events for the other two buttons. First, double-click the Update Score button in the MainPage designer. This creates the btnUpdateScore_Click method in the code page. Change the method so that it looks like the following:

C#
```
private async void btnUpdateScore_Click(object sender, RoutedEventArgs e)
{
    currentScore.Score = Convert.ToInt32(tbxScore.Text);
```

```
    await App.MobileService.GetTable<GameScore>().UpdateAsync(currentScore);
    refreshList();
}
```

VB.NET

```
Private Async Sub btnUpdateScore_Click(sender As Object, e As RoutedEventArgs)
    currentScore.Score = Convert.ToInt32(tbxScore.Text)
    await App.MobileService.GetTable(Of GameScore)().UpdateAsync(currentScore)
    refreshList()
End Sub
```

Do the same for the Click event on the Delete Score button. Again, in the designer for MainPage, double-click the button. This generates the Click event handler, which you should modify to look like the following:

C#

```
private async void btnDeleteScore_Click(object sender, RoutedEventArgs e)
{
    await App.MobileService.GetTable<GameScore>().DeleteAsync(currentScore);
    refreshList();
}
```

VB.NET

```
Private Async Sub btnDeleteScore_Click(sender As Object, e As RoutedEventArgs)
    await App.MobileService.GetTable(Of GameScore)().DeleteAsync(currentScore)
    refreshList()
End Sub
```

The final piece of code is not so much about WAMS as it is about making the application function the way you expect it to. If you look at the code for MainPage, you see a method named OnNavigatedTo. This method is basically invoked when the page is reached. In this event, you would like the list of game scores to be populated. As a result, add a call to the refreshList method to the body of this method.

Now that your coding is complete, launch the application by pressing F5. Put an integer value into the Score text box and click the Add Score button. You should see the score appear in the list box. Change the value in the Score text box and click Add Score again. Now there should be two scores. Change the value in the Score text box once again and click Update Score. The number of scores stays the same, but the bottom one has the updated score. Finally, click the Delete Score button and notice that you're now down to one score (see Figure 2-10).

The Data Model

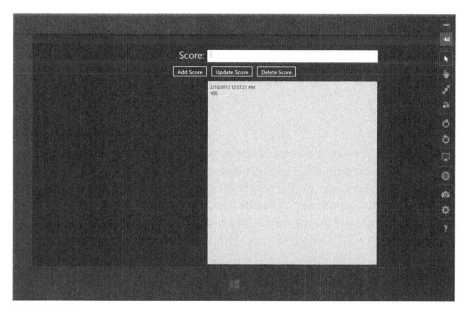

FIGURE 2-10

Error Handling

The code to this point has been what is called "happy path" code. It presumes that nothing can go wrong with the mobile service. Although that's nice for code that is in a book, it's not good enough in the real world, because communications with WAMS is network code. It goes out through the Internet to talk to your mobile service. When you depend upon network code for your application to function, you need to code for the possibility of failure.

The thought behind this is that, when your application is out in the wild, you have no idea about the network connectivity that is available for your application. The user could be running in airplane mode while travelling. Your user's network could be clogged by someone in their house streaming the latest episode of *Game of Thrones*. The cause of the problem is irrelevant, but operating in a non-Internet mode is a test case that is part of Microsoft's application certification, so you might want to consider how your app will handle it.

When it comes to surfacing issues through the WAMS client proxy, you will notice that none of the methods have a return value. This means that your only way to be notified of an issue is through standard exception handling. Consider it a requirement that all of your calls to WAMS be wrapped in a try-catch block.

You should be aware, however, that the exception thrown might not always be the result of network issues. In Chapter 3, you look at how to use the server-side scripting capabilities of WAMS to detect problems and terminate the requested operation. In doing so, it sets the status code of the HTTP response. This appears to the client code as a `MobileServiceInvalidOperationException`. So make sure that you look for both standard network exceptions and this mobile service exception when you are making your code bullet proof.

SUMMARY

Fundamentally, WAMS is about storing and retrieving data. And as you have seen in this chapter, the techniques for doing this are very easy to integrate into any application that you're building. In the next chapter, we move from the client side of the functionality to the scripting that can take place on the server-side. So while you will be using these client-side functions over and over again, it's really on the server-side where the possibilities start to expand.

3
Mobile Services Validation

IN THIS CHAPTER:

- ➤ Understand the basic structure of WAMS server-side scripting
- ➤ Learn about the properties of the User object that can be put to use in the scripts
- ➤ Update the WAMS database directly using the mssql object

WROX.COM CODE DOWNLOADS FOR THIS CHAPTER

The wrox.com code downloads for this chapter are found at http://www.wrox.com/go/windowsazuremobileservices on the Download Code tab. The code is in the Chapter 3 download and individually named according to the names throughout the chapter.

In the previous chapters, you learned how to create an application that utilizes Windows Azure Mobile Services (WAMS) to perform simple CRUD functionality. Getting to this point is fundamental to almost every business application. However, creating and updating data is not sufficient. To ensure that the data related to your application maintains its integrity, you need to perform some level of validation.

Of course, you could reasonably expect that the user interface component will do some of this validation work, but that will not be sufficient. To understand why, let's consider, for a moment, data validation in a more common scenario: web applications.

It is considered a best practice in web applications that you should not accept any information that comes from across the Internet without subjecting it to further scrutiny. This is true even if the web page that submits the data has all sorts of validation. It is very easy for someone to create an HTTP request that mimics a web page without submitting the data to validation.

Access to the data in a mobile service is done across the Internet. This is true whether you are using the `MobileServiceClient` object or the REST interface. So regardless of how much validation you have implemented on the client, the same validation should be done on the server.

For WAMS, server-side validation comes in the form of scripts written in JavaScript. This chapter takes an in-depth look at these scripts, and provides several examples of common usage.

ADDING SERVER-SIDE SCRIPTS

Naturally, the starting point in the discussion of WAMS scripts is how to access them. Scripts are defined at the table-level, which means that each table can have a different set of scripts. To start, go to the dashboard for your mobile service (Figure 3-1).

FIGURE 3-1

The Data link at the top gives you access to the tables that are part of your mobile service. Click that link to display the list of tables (Figure 3-2).

FIGURE 3-2

Adding Server-Side Scripts

Next, click the table for which you are creating scripts. This gets you to the table screen, shown in Figure 3-3.

At the top of the table screen are a number of links related to the table, including the Script link. Clicking that link takes you to the page that is used to define the scripts for the current table, as shown in Figure 3-4.

FIGURE 3-3

FIGURE 3-4

You can create four different scripts for each table — Insert, Update, Delete, and Read — which correspond to the four CRUD operations. The script for deleting is actually called `del` because `delete` is a reserved word in JavaScript, but you get the idea.

Selecting the script that you want to work with is easy. The four choices appear in a drop-down control (the window that appears for the Insert option is shown in Figure 3-4). Choose the desired operation and a code editing window appears that contains the current method. The default method is the functional equivalent of a no-op instruction.

The code editor window is not particularly sophisticated, as far as editors go. It has keyword coloring and line numbers and not much else; so don't look for IntelliSense here.

Regardless of the operation, each method takes three parameters. The first parameter differs based on the operation, but the second and third are consistent across each of the operations. The second is the user information for the authenticated user, and the third is an object that represents the request itself. You can find more details about how they are used in this and subsequent chapters.

Inserting Data

The `insert` function is invoked when a record is being inserted into the data store. The signature for this function includes the user information object, the request object, and the item being inserted. The default method is shown here:

```
function insert(item, user, request) {
    request.execute();
}
```

Note the call to the `execute` method on the request object. It is this call, regardless of the operation, that actually completes the request and performs the operation. If you wrote a method that did not call `execute`, the requested function would not be performed.

Naturally, this leads to the question of how the execution of a particular function can be prevented. Consider the `insert` method shown here:

```
function insert(item, user, request) {
    if (!item.approved) {
        request.respond(statusCodes.FORBIDDEN, "You can only insert approved items.");
    } else {
        request.execute();
    }
}
```

First off, you see the introduction of some logic into the `insert` method's script function. This logic is illustrative of how validation is performed within WAMS. In this particular case, the item being inserted has a Boolean property named `approved`. If the value of the property is `true`, the insert is performed. However, if the value is not `true`, the response to the request will be an HTTP status code 403 (that's what the FORBIDDEN status code translates to), along with an error message.

Please take note that if you create a method that does not perform either an execute or a respond through all of the code paths, you might create a situation in which the mobile service does not return a response to the request. You will see this behavior manifest itself as a request timeout on the requesting side of the exchange.

Updating Data

The update function is invoked when an existing record in the data store is being updated. The signature for the method is the same as the insert method, in that the first parameter is the item being updated, and the other two are the requesting user and the request. The default method is as follows:

```
function update(item, user, request) {
    request.execute();
}
```

WAMS allows the item that is being updated (or inserted for that matter) to be modified as part of the script's processing. For instance, a field such as lastUpdated could be given the current time automatically. The result would be something that looked like the following:

```
function update(item, user, request) {
    item.lastUpdated = new Date();
    request.execute();
}
```

What is even more interesting is that the lastUpdated field does not need to exist on the object being updated prior to running the script. Go back to one of the examples from Chapter 2 for a moment. Following is the class definition for the object that is being saved:

C#

```
[DataTable(Name="gameResult")]
public class GameScore {
   [DataMember(Name="id")]
   public int Id { get; set; }

   [DataMember(Name="score")]
   public int Score { get; set; }

   [DataMember(Name="resultDate")]
   public DateTime GameDate { get; set; }
}
```

VB.NET

```
<DataTable(Name="gameResult")> _
Public Class GameScore
```

```
    <DataMember(Name="id")> _
    Public Property Id As Integer

    <DataMember(Name="score")> _
    Public Property Score As Integer

    <DataMember(Name="resultDate")> _
    Public Property GameDate As DateTime
End Class
```

Based on this definition, there is no property called `lastUpdated`, nor any property that is mapped into `lastUpdated`. Yet if this object were passed into a WAMS table that used the preceding script, there would be no problem. The field would be added to the persisted data record and could then be used as part of any other script. In other words, you can dynamically add fields to a data record as part of the script processing, including fields that would only ever be used by WAMS.

Deleting Data

The `del` function is invoked when a record is in the process of being deleted. The execution of the method takes place prior to the deletion, so it is possible to cancel the process instead of finishing the removal. The signature for this function includes the user information object, the request object, and the ID of the item being deleted. Notice that the signature doesn't include the item itself. (The technique to retrieve the item that is about to be deleted is discussed a little later.) The default method is shown here:

```
function del(id, user, request) {
    request.execute();
}
```

The mechanism for canceling the deletion is not nearly as complicated as it sounds. Recall from the "Inserting Data" section that the call to the `execute` method is what actually triggers the function (the deletion in this case) to take place. The implication for canceling the deletion is that if you simply don't call `execute`, the deletion is avoided. This is largely the case, but it is not the entire story.

You might also recall from the previous section that you need to call either the `execute` method or the `respond` method, and if you are canceling the deletion, the `execute` method isn't being called. That leaves the `respond` method.

The first parameter in the `respond` method is the error code to return to the client. If you are canceling the deletion but don't want the client to see the cancellation as an error, then pass `statusCodes.OK`. If you would like the client to be informed of the cancellation, then pass `statusCodes.FORBIDDEN` or `statusCodes.UNAUTHORIZED`, depending on what information you are trying to convey back to the client.

As was mentioned at the beginning of the section, only the ID for the record is passed into the method. If you want to use information in the record to control whether the deletion is allowed, you first need to retrieve the record from the data store. You can find an example of how to perform a query against the data in a table in the "Storing Per-User Data" section later in this chapter.

Retrieving Data

The `retrieve` function is invoked when a query is made that retrieves one or more records from the data store. The execution of the method takes place prior to the actual query being run against the data. As a result, the typical use case is to add criteria to the query to additionally limit the records that are returned. The default method is shown here:

```
function read(query, user, request) {
    request.execute();
}
```

The signature for the `read` method is a little different than the other three methods. Fundamentally, the parameters fill the same purpose, and the `user` and `request` parameters are the same as with the other methods. However, the first parameter, `query`, contains the query that is being used to retrieve the data (as opposed to the data record or the ID).

One of the primary use cases for placing code into the `read` function is to add a filter to each of the requests being made. For example, if the data in your mobile service has been collected on a per-user basis, any query should return only the records for the current user. So consider for a moment on how that can be done.

The `query` parameter is actually a `Query` object that represents the filters, predicates, ordering, and projections that are being executed against the data store. To provide access to these properties, the `Query` object has a number of methods that can be used. For this particular example, the `where` method fits our needs.

The `where` method actually has two overloads. One of them takes a JavaScript object where the property names match columns in the data store, and the values in the object are compared to values in the data store. The second takes a JavaScript function that is converted to a `where` clause when the query is processed.

The first overload, when limiting the records in the data store to those that belong to the current user, produces a function that looks like the following:

```
function read(query, user, request) {
   query.where({ userId: user.userId });
   request.execute();
}
```

In this example, the `userId` property on the user object that has been passed into the method is used as the value in the JavaScript object. The name of the property (and therefore the name of the column in the data store) is `userId`.

It is important at this point to realize that JavaScript is a case-sensitive language, even to the point where the case of the property in the JavaScript object must match the column in the data store in order for the comparison to be done. This also means that the property added way back in the `insert` function also needs to have the same case.

Making sure the case is consistent is simple, and the next step is easy as well. Consider the following snippet:

```
function read(query, user, request) {
   query.where({ userId: user.userId, isApproved: true});
   request.execute();
}
```

Instead of an object with a single property passed into the `where` method, the object is a little (but only slightly) more complicated. With this example, the filtering that is applied is additive, which means that the records retrieved from the data store will be those for which the user ID is the same as the current user's ID, and the `isApproved` value is `true`. Just to make sure that we're on the same page, it would be like writing the following SQL statement:

```
SELECT * FROM TableName
WHERE userId = user.userId AND isApproved = true
```

Although it might not be obvious, the criteria that you're adding are also added to any criteria that were originally in the query object. As was demonstrated in the Retrieving Data section in Chapter 2 - Creating and Manipulating Data, LINQ can be used to specify criteria that is then sent to the mobile service. The criteria that are added in the server-side script extend the criteria provided by the client.

Finally, and for completeness, you could have written the query using one of the following examples and achieved the same result:

```
function read(query, user, request) {
   query.where({ userId: user.userId});
   query.where({ isApproved: true});
   request.execute();
}

function read(query, user, request) {
   query.where({ userId: user.userId})
        .where({ isApproved: true});
   request.execute();
}
```

The second overload for the `where` clause takes a JavaScript function as a parameter. So to accomplish the same functionality as the first method, the `read` function would look like the following:

```
function read(query, user, request) {
   query.where(function(currentUserId){
```

```
        return this.userId == currentUserId;
}, user.userId));
    request.execute();
}
```

You need to be aware of a couple of points with regard to the JavaScript function. First of all, notice that the only statement in the body of the function is a return. This (at least at the moment) would appear to be a requirement. If you have more than one statement in the body, you will receive an exception raised in the client. To give you a sense of what this looks like, consider the following snippet:

```
function read(query, user, request) {
    query.where(function(){
        int i = 500;
        return this.highScore > i;
}));
    request.execute();
}
```

In this case, it would appear that only those records with a `highScore` value greater than 500 will be returned. However, as already mentioned, an exception would be received by the client. The reason for this behavior relates to how the function is actually processed. Unlike what your intuition might suggest, the function is *not* run after the records have been retrieved. In fact, if you think about it, you really don't want WAMS to do this for performance reasons. What you would actually like to happen is to have the function translated into criteria that is then executed directly on the server, and that is what happens.

The expression that is in the return statement is translated into a WHERE clause and submitted to the database server. The filtered result set is then returned to the client where it is processed. No additional server-side functionality is added as a result of using the JavaScript function (good for performance). But there is also no way to define a function that is truly executed each time a record is processed to determine if it should be part of the result set.

The User Object

Along with being able to set permissions on the table CRUD functions through the Azure portal, WAMS also provides a mechanism that allows branching in the server-side scripts based on the current user and the level of that user. To get a sense of what's possible, this section looks at the User object that is passed into each one of the methods.

The purpose of the User object is to represent the user who is making the request. The amount of information that is available through the object is sparse, to say the least. There is a userId property, which contains the user ID of the user if he or she has been authenticated. If the user has not been authenticated, the value of the userId property is undefined. The different mechanisms to authenticate users are covered in Chapter 4.

The `User` object also has a `level` property. Again, this value is dependent upon whether the user has been authenticated. If there has been no authentication, the level is set to `anonymous`. For authenticated users, the value is either `authenticated` or `master`. The value of `master` is determined by the presence of a master key in the request. More specifically, the master key for your mobile service is located on the same screen as the authentication key.

Finally, there is an `accessTokens` property that contains the access token returned by the authentication provider. The property itself is a JSON object that has the following format:

`{ providerName: "access_token" }`

The `providerName` indicates the service that performed the authentication. The format and range of values for the access token depend on how the authentication was performed and, depending on the provider, could be used to access additional functionality from the server-side scripts.

Common Scenarios

As has already been noted, the server-side scripts are written in JavaScript. It seems like overkill (not to mention tedious) to provide a complete reference to the JavaScript functions that you are able to use. Syntactically, it supports the basic JavaScript command and operations. However, some additional objects are worth covering and, in this context, it is worthwhile to look at some very common scenarios and provide examples of how they might be implemented. At the same time, this will provide a platform to describe the other WAMS-specific objects that are available for you to use.

Storing Per-User Data

It's arguable whether the need to store data for individual users is the most common use for server-side scripts, but there is little question that it is a very frequently requested piece of functionality. Before showing the scripts, it's important to think about what the requirement really is.

To start, the idea is to allow users to perform CRUD functions on data that "belongs" to them. Typically, that means that when a retrieval is executed, the only records in the result will be those associated with the current user. It would be easy to have the client perform the filtering request, but that is not the best solution, at least from a security perspective. The reason is that if your mobile service depends on the fact that a client application is retrieving only the data related to the current user, how would the mobile service defend itself from a nefarious person who makes a REST-based request for data without including the filter? The answer is: it can't. For that reason, using client-side filtering is not an appropriate solution.

Because client-side filtering will not work, you must turn to the server side. The starting point is to change the permission on the target table so that only authenticated users can access the table. This process is explained in "The Data Model" section in Chapter 2 - Creating and

Adding Server-Side Scripts

Manipulating Data, and the mechanism of actually authenticating users is described in Chapter 4 - The Authentication Options in WAMS.

Once authentication is required on the table, it's time for you to tackle the scripts. What the scripts need to do can be broken down by the four CRUD functions:

- **Create** — Automatically assign the current user's ID to the record that is being created.
- **Retrieve** — Add a filter to the query so that only information for the current user is returned.
- **Update** — Check the record being updated to see if it belongs to the current user prior to it being updated.
- **Delete** — Check the record being deleted to see if it belongs to the current user prior to it being deleted.

With this specification in hand, you can code the server-side scripts to implement it. Start by opening the insert script for the table. As mentioned, the script needs to assign the current user ID to the record. The following script does just that:

```
function insert(item, user, request) {
    item.UserId = user.userId;
    request.execute();
}
```

You should take note of two things in this code. First, because the script is accessible only to authenticated users, the userId property of the User object is guaranteed to have a value. Second, the value of the userId is assigned to a property on the item being created; in particular, to the UserId property. The important part is that there is no need for the UserId property to exist on the incoming object prior to the insert function being invoked. In other words, do not feel that you need to create a UserId property on the data transfer object that is sent to your mobile service. The dynamic nature of JavaScript, combined with the dynamic schema generation in WAMS, will ensure that the user ID is not only associated with the object, but will also be stored in your mobile service's database.

Now that the basic mechanism for tracking data on a per-user basis has been determined, you need to apply it to the other functions. Start with the retrieval, which is the easiest to implement:

```
function read(query, user, request) {
    query.where({ UserId: user.userId });
    request.execute();
}
```

For the retrieval you want to apply the user ID filter to every request that is made. You do this through the where method. Fortunately, the filtering applied by the where method is cumulative. This means that even if the retrieval already has criteria associated with it, the user ID filter will be added to it.

The update and delete functions are similar to one another, but a little trickier to implement. Consider that in both cases, the ID property in the incoming object is used to determine the target object to modify. There is no guarantee that the ID being passed in represents a record that belongs to the user making the request. Keep in mind that a request can be made using REST from an authenticated user and that request can easily be constructed to contain an arbitrary ID. So for both of these functions, you need to take the incoming ID, query the data store for that object, and determine if the userId for that object matches the userId of the requestor. The following script accomplishes this for the update function:

```
function update(item, user, request) {
    var table = tables.getTable('gameResult');
    table.where({ id: item.id }).read({
        success: function (results) {
            if (results.length) {
                var existingItem = results[0];
                if (existingItem.UserId === user.userId) {
                    request.execute();
                } else {
                    request.respond(statusCodes.BAD_REQUEST, "Invalid user");
                }
            } else {
                request.respond(statusCodes.NOT_FOUND);
            }
        }, error: function () {
            request.respond(statusCodes.NOT_FOUND);
        }
    });
}
```

You can glean a number of useful lessons about WAMS scripting from this script.

There is a built-in tables object that provides a method (getTable) to retrieve information about a specific table. This script retrieves the gameResult table, because that is the table with which this script is associated. Unfortunately, the table name needs to be hard-coded, but that is less of an issue because WAMS does not allow for scripts to be shared across different tables.

Once you have retrieved the desired table, use the where method to filter on the ID property, then invoke the read method to retrieve all of the matches. The read method takes two parameters: one is the JavaScript function to execute upon successful completion, and the other is the JavaScript function to execute if the read method causes an error.

The error portion of the process is easy. If there is a problem retrieving the data, an error with a status code of Not Found (HTTP 404) is returned to the client.

The success portion is a little more complicated, although only a little. First, the number of records that are returned is checked to ensure that there is at least one. If no records were found that matched the user ID, an HTTP 404 - Not Found is returned. For the first record that is returned, check to see if the user ID matches the current user. If not, then return the bad request status code (HTTP 400) along with the "Invalid user" message. If both of these checks pass without incident, then perform the update.

Once you understand the logic found in the update function, you'll realize that the script to handle deleting a record is almost identical. The differences between the two cases are that the name of the function has changed (del vs. update) and the fact that the ID of the object to be deleted is passed into the method, not the object itself. The following example shows the del function:

```
function del(id, user, request) {
    var table = tables.getTable('gameResult');
    table.where({ id: id }).read({
        success: function (results) {
            if (results.length) {
                var existingItem = results[0];
                if (existingItem.UserId === user.userId) {
                    request.execute();
                } else {
                    request.respond(statusCodes.BAD_REQUEST, "Invalid user");
                }
            } else {
                request.respond(statusCodes.NOT_FOUND);
            }
        }, error: function () {
            request.respond(statusCodes.NOT_FOUND);
        }
    });
}
```

At this point, the server-side is complete. What is even more compelling than the underlying ease of the implementation is the fact that your client side didn't have to change at all. You can add this functionality without even needing to redeploy a new version of the client application. That is the sort of flexibility that is very common throughout WAMS.

Inserting Multiple Items per Call

The default behavior for WAMS is to act on one record with each call. It's true that a call to WAMS involves a network, with all of the potential latency and failure issues that it entails. Still, for most scenarios, the one-to-one relationship is quite acceptable.

However, situations arise for which it is desirable to act on more than one record per call. Possible reasons include an extreme desire to minimize network traffic or the need to bundle the updates into a single atomic transaction. (Yes…that's right. Multiple calls to WAMS do not participate in a single transaction.) Some techniques, however, enable you to do this. It requires a little more effort from the developer, but it is not beyond the capability of anyone reading this book.

Like the previous scenario, look at the functionality required in each of the CRUD functions:

- **Create** — In the object that is passed to the `create` method, loop through the items that are to be created with this single call and create the records in the data store.
- **Retrieve** — No functionality needs to be added. The ability to update multiple records does not impact the retrieval of the items.
- **Update** — In the object that is passed to the `update` method, loop through the items that are to be updated and modify the corresponding records in the data store.
- **Delete** — In the object that is passed to the `delete` method, loop through the items that are to be deleted and remove the corresponding records in the data store.

You'll notice two things about these specifications. First, the Retrieve function requires no change at all. The second is that the basic processing for the Insert, Update, and Delete functions are identical: loop through the elements to be processed and process them. The creation step is probably the most straightforward, so start with that by considering the following script:

```
function insert(item, user, request) {
    var table = tables.getTable('gameResult');
    insertScores(table, request, item.scores);
}
```

This script isn't doing that much. It retrieves a reference to the `gameResult` table. It then passes that table, along with the request and the items to be inserted, into the `insertScores` method. But that does bring up an interesting point. As you create server-side scripts for WAMS, you can also create additional methods to simplify the scripting. This is not to say that methods that you add to the insert script can be used in other functions (they can't), but it can clean up the code that you're writing:

```
function insertScores(table, request, scores) {
    var index = 0;
    scores.forEach(changeGameDate);
    var insertNext = function () {
        if (index >= scores.length) {
            request.respond(201, { id: 1, status: 'inserted successfully' });
        } else {
            var scoreToInsert = scores[index];

            table.insert(scoreToInsert, {
                success: function () {
                    index++;
                    insertNext();
                }
            });
        }
    }
```

```
    };

    insertNext();
}
```

Walking through this script reveals a number of interesting components of WAMS scripting. First, you see a statement where the changeGameDate method is invoked for each of the items in the list. The reason for this is covered shortly.

Second, you see that a function named insertNext is created. The purpose for creating this function is not readily apparent. It would seem more logical to simply enter into a for loop and call the insert method on the table object for each score in the list. However, the insert method is actually an asynchronous one. If you just loop through the list, calling insert each time, you have no control over the ordering or threading that goes on, and it will be very complicated to track when all of the items have been inserted (so that you can return a message to the client). So instead, after each insertion completes, the same function (insertNext) is invoked to insert the next item in the list. It takes a little longer, but gives you more control over the result with less complexity.

By the way, this script is almost identical to the Update or Delete functions. The difference is that instead of invoking the insert method on the table object, you would invoke the update or delete method.

Finally, back to the changeGameDate method and the reason for its existence — its existence and utility applies only if you are using the dynamic schema capability in WAMS.

First off, please keep in mind that there is no such a thing as a "JSON date." In JSON, dates and times are sent over the wire as a string. WAMS handles dates by looking at the string and, if it has a specific format (that format is ISO 8601, for those keeping track of the details), the value is converted to a JavaScript Date object. However, for reasons of performance, this conversion is only done in top-level objects. There is no traversal of the entire object graph looking for dates to convert. As a result, the GameDate property in the GameScore class is not converted to a date within WAMS. If that column (the GameDate column in this example) does not exist in the data store when the insertion is performed, the column will not be given the appropriate data type. The created column will be a string (technically, an nvarchar(max) data type), not a date.

To address this, before any of the data is inserted, you need to convert the GameDate property in the incoming objects from a string type to a Date type. The changeGameDate function shown here accomplishes this:

```
function changeGameDate(obj) {
    var gameDate = obj.GameDate;
    if (typeof gameDate === 'string') {
        gameDate = new Date(gameDate);
        obj.GameDate = gameDate;
    }
}
```

The one item that is left to include in this example is the class that is passed to WAMS. Following is the declaration:

C#
```
[DataTable(Name="gameResults")]
public class GameScores {
   [DataMember(Name="id")]
   public int Id { get; set; }

   [DataMember(Name="scores")]
   [DataMemberJsonConverter(ConverterType = typeof(ScoreListConverter))]
   public List<GameScore> Scores { get; set; }
}
```

VB.NET
```
<DataTable(Name="gameResults")> _
Public Class GameScores
   <DataMember(Name="id")> _
   Public Property Id As Integer

   <DataMember(Name="scores")> _
   <DataMemberJsonConverter(ConverterType = typeof(ScoreListConverter))> _
   Public Property Scores As List(Of GameScore)
End Class
```

You can find the GameScore object referenced in this code in Chapter 2. Note that the Name property in the DataTable attribute needs to match the table with which your script is associated. It does not need to be the name of the table in which the data will ultimately reside.

The property that contains the list of objects to be stored (the Scores property) is decorated with the DataMemberJsonConverter attribute. This attribute indicates the class that will be used to convert the property value into a JSON array. In this case, the ScoreListConverter class is used. The implementation of this class is shown in the following code:

C#
```
public class ScoreListConverter : IDataMemberJsonConverter
{
    public object ConvertFromJson(IJsonValue value)
    {
        return null;
    }

    public IJsonValue ConvertToJson(object instance)
    {
        JsonArray result = new JsonArray();
        IEnumerable<GameScore> scores =
```

```
            (IEnumerable<GameScore>)instance;

        foreach (var score in scores)
        {
            result.Add(MobileServiceTableSerializer.Serialize(score));
        }

        return result;
    }
}
```

VB.NET

```
Public Class ScoreListConverter
    Implements DataMemberJsonConverter

    Public Function ConvertFromJson(value As IJsonValue) As Object

        Return Nothing

    End Function

    Public Function ConvertToJson(instance As Object) As IJsonValue

        Dim result = New JsonArray()
        Dim scores As IEnumerable(Of GameScore) = _
            CType(instance, IEnumerable(Of GameScore))

        Dim score As IEnumerable(Of GameScore)
        For Each score In scores
            result.Add(MobileServiceTableSerializer.Serialize(score))
        Next

        Return result

    End Function
End Class
```

The purpose of this converter is to take the list of `GameScore` objects and convert them into a JSON array. The conversion of objects is done through the `MobileServiceTableSerializer`, which builds up a JSON object that WAMS will recognize. Naturally, you could create it yourself, but if you plan on doing that, it is recommended that you look at the object that gets transferred across the wire to make sure that you send an object that WAMS can use.

One of the elements of this functionality discussed earlier was the possibility of having all of these updates take place inside of an atomic transaction. The script that you have written to this point is definitely not that. In fact, each invocation of the `execute` method is independent of one another and each is subject to pass or fail as the whims of the database deities determine. And, if you look at how the application is constructed, if you were modifying ten items and an error took place on the fifth one, you would have four modified items and six that remain unchanged.

If atomicity is your goal, a different approach is required. In the world of WAMS, this approach utilizes the mssql object. The purpose of the mssql object is to allow the WAMS scripts to gain access to the full power of T-SQL. In WAMS, mssql is an intrinsic object. There is no need to instantiate it. There is also no need to specify any connection information for the object. The object "knows" about the database that is associated with the mobile service, and so no additional information is required.

The mssql object has a query method that is used to execute SQL commands against the database. The basic form of this method is as follows:

```
mssql.query(sqlStmt, positionalParameterArray, {
    success: function () {
        doSuccessStuff();
    },
    error: function () {
        doErrorStuff();
    }
});
```

The first parameter is the SQL statement that is to be executed. The allowable syntax is pretty much exactly what you'd expect out of T-SQL. The underlying database is (basically) a SQL Server, after all. However, because it is actually a SQL Azure database, it does have the same limitations that you have if you write T-SQL against the database directly. The vast majority of queries you are likely to write will work, but some of the commands and options are not available due to the shared and virtualized nature of the environment.

The second parameter is a JavaScript array of positional parameters. And yes, you read that correctly. Although you can specify parameters in the SQL statement, they are positional and not named. For instance, the following SQL statement accepts two positional parameters:

```
INSERT INTO gameResult (Score, GameDate) VALUES (?, ?)
```

As a result, the code to execute this statement using the mssql object would be as follows:

```
mssql.query('INSERT INTO gameResult (Score, GameDate) VALUES (?, ?)',
    [500, '3/1/12'], {
    success: function () {
        doSuccessStuff();
    },
    error: function () {
        doErrorStuff();
    }
});
```

Armed with this information, you can modify the various scripts to make the changes atomic. The following script is for inserting the records, but you can easily modify the SQL that is created in order to perform updating or deleting atomically as well:

```
function insert(item, user, request) {
    var sqlStmt = '';
    item.scores.forEach(function (game) {
        sqlStmt = sqlStmt + 'INSERT INTO gameResult (score, gameDate) ' +
            'VALUES (' + game.score + "', '" + game.gameDate + "');";
    });
    mssql.query(sqlStmt, [], {
        success: function () {
            request.respond(201, { id: 1, status: 'inserted successfully' });
        },
        error: function () {
            request.respond(statusCodes.BAD_REQUEST, "insertion failed");
        }});
}
```

The flow of the script is to build up the SQL statement and then execute it. Because it is possible to create a batch of SQL statements by separating each statement with a semicolon, ultimately you will have string of SQL statements, each of which inserts a record and yet, when executed through the mssql object, performs the combination of functions atomically.

SUMMARY

The ability to run scripts on the server side, triggered by CRUD functionality is a very power option. It allows you to perform validation on incoming data, as well as implement filtering and some more complex data relationships. But, as you might guess, these are not the only script-based scenarios that you will find in this book. In fact, many of the advanced features of WAMS are implemented through scripts of one form or another. Yet through all of those scripts, the basics that you have just learned will continue to hold.

4 Authentication Options in WAMS

IN THIS CHAPTER:

➤ Configure your mobile service to support authentication

➤ Set up the various authentication providers to support your mobile service

➤ Troubleshoot issues related to authentication and your mobile service

WROX.COM CODE DOWNLOADS FOR THIS CHAPTER

The wrox.com code downloads for this chapter are found at http://www.wrox.com/go/windowsazuremobileservices on the Download Code tab. The code is in the Chapter 4 download and individually named according to the names throughout the chapter.

To this point, you've seen how to access your mobile service from the client side, as well as how to implement server-side scripts to handle validation and other advanced functionality. However, to take advantage of these capabilities, in many cases, you also need to be able to identify the user who is making the request.

As of the time of writing, Windows Azure Mobile Services (WAMS) supports authentication from four different providers: Facebook, Google, Microsoft Account (formerly Microsoft Live ID), and Twitter. From the perspective of creating client-side code that performs this authentication, the steps are very straightforward. This chapter covers a number of topics related to authentication. First, you learn some of the fundamental ideas related to authentication. This is not because you need to have this information to utilize WAMS authentication, but because when something goes wrong in the authentication process, it's a very good idea to have this background knowledge. If you're already familiar with federated identity, feel free to skip to the second section.

The second section covers the four different providers. This includes the steps you need to go through to properly register your application so that it can support WAMS authentication. Finally, the third section shows you the client-side code that is necessary to authenticate a user.

Federated Authentication

One of the key ideas behind *federated authentication* is trust. More specifically, how much trust two, potentially unrelated business organizations have for one another. The idea behind any federated authentication protocol is to create a mechanism that allows user to provide their credentials to one domain and then allow the authentication of those credentials to be accepted as valid by another domain. You might also hear the term *single sign-on* applied to this process as well, as the need for users to create and managed duplicate accounts across different applications and domains is removed. To get an understanding of why trust matters, consider the scenario illustrated in Figure 4-1.

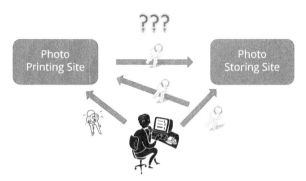

FIGURE 4-1

The scenario illustrated in Figure 4-1 is one of a number of similar patterns. You, as a user, have created an account on a website (a photo-sharing site, such as Flickr). Naturally, you have defined your credentials, and because the account is yours, you have full control over anything that can be done on that website. In the case of a photo-sharing site, this might mean uploading photos, tagging or commenting on the photos, or even deleting them. In addition to you being able to access your "source" site, you would like another site (the photo-printing site) to be able to view your photos.

So how do you allow this? Well, the obvious (and much less than optimal) approach is to give your Flickr credentials to the photo-printing site. Now the site has no problem accessing the photos. Of course, if the site was in any way malevolent, it would also have no problem updating, tagging, deleting, or performing other actions on any of your photos as well. To the photo-sharing site, any request coming from the photo-printing site is indistinguishable from a request coming from you personally. Not a good situation. So, consider a more appropriate approach, as illustrated in Figure 4-2.

This involves a trusted third-party (the STS Provider in the figure). You need to trust this third party to keep your credentials and not use them for any nefarious purposes. The photo-printing and photo-sharing site need to trust the third party as being capable of identifying users who have access to the site. In other words, instead of maintaining their own database of credentials, each of these sites uses the authentication services of the STS provider.

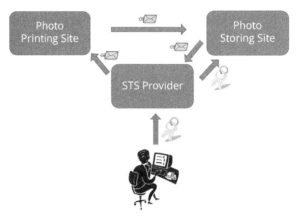

FIGURE 4-2

Here's a walkthrough of the flow that is illustrated in Figure 4-2. As part of logging in to the photo-printing service, you identify yourself to the third party. In return, the photo printing service gets get a token. This token contains not only information about your identity (your user ID, for example), but also (optionally) a set of permissions that you're willing to grant to the photo printing service (in this case, to read your photos).

Now when the photo-printing service wants to gain access to your photos, the service presents the token to the photo-sharing service. Because the photo-sharing service has trusted that the third party is capable of authenticating users, any user information returned to the service from the third party is considered valid. Therefore, the credentials and permissions related to the token can now be used to satisfy the requests from the photo-printing service.

The third party in this story is an example of a *Secure Token Service* (STS). A number of STS providers are available, including Facebook, Google, Microsoft Account, and Twitter. In addition, Windows Azure has a service known as Access Control Service (ACS) that provides this functionality as well. At the moment, ACS is not supported for mobile services, but who knows what the future will bring.

So let's apply this to your situation. You have an application (the platform on which it is running is not important) that would like to communicate with WAMS. Your user does not want to trust you with his credentials. So your application provides a mechanism that allows the user to log in to the STS provider. The user then gives you a token, which is presented to WAMS. WAMS uses that token to retrieve the user information from the STS, and the user's identity is now available to your server-side scripts.

Given the element of trust that is required, creating your own authentication mechanism is pretty much out of the question. WAMS, as it turns out, supports the four STS providers that were previously mentioned (Facebook, Google, Microsoft Account, and Twitter).

Naturally, the specifics of this process are a little more complicated. It involves a number of different conversations that take place between your applications, the STS provider, and your mobile service. But the basic flow is as described. And, even better, the underlying details of the conversation are hidden from your view. As you will see in the section describing the client side of the process, little more than a single line of code is required to initiate the authentication. For your mobile service, the entire conversation takes place out of view. You see the `userId` property in the `User` object that is passed into the server-side scripts without needing to be concerned about how that information was made available.

I need to explain one additional piece of functionality to close the loop on OAuth. It allows some of the configuration of the SRS providers that you're about to do to make sense.

As mentioned earlier, both the photo-printing and photo-sharing services need to be configured to interact with the STS provider. The configuration involves the exchange of an identifier and a shared secret. The identifier allows the STS provider to uniquely recognize the requesting service, and the shared secret allows communication between services and the STS provider to be secured. For your application to take advantage of these providers, you need to register your application with them to get the application ID and secret information. That process is covered in the next section.

Setting Up the Authentication Providers

As already mentioned, WAMS supports four different authentication providers. For your application to utilize a provider, you need to configure the shared key that was described in the previous section. The next few sections cover the steps you need to go through for each provider so that your application is able to take advantage of authentication.

Facebook

To use Facebook as an authentication provider, you must be registered as a developer and you also need to have a Facebook account. If you don't have a Facebook account, visit `http://www.facebook.com` and create one. Then, once it's ready, visit `http://developer.facebook.com` (see Figure 4-3).

If you have not previously registered as a Facebook developer, you'll see a Register Now button in the top-right corner of the screen. Click it and follow the directions to register as a developer. If you are already a developer (or once you have registered), click the Apps link at the right side of the top menu. The screen where you can manage your Facebook apps appears (Figure 4-4).

CHAPTER 4 AUTHENTICATION OPTIONS IN WAMS

FIGURE 4-3

FIGURE 4-4

Authentication Options in WAMS

If you have existing apps (as you can see in Figure 4-4), they are visible and you can edit the information at this point. However, you're creating a new application in this process so click the Create New App button in the top right of the screen. A dialog box appears, as shown in Figure 4-5.

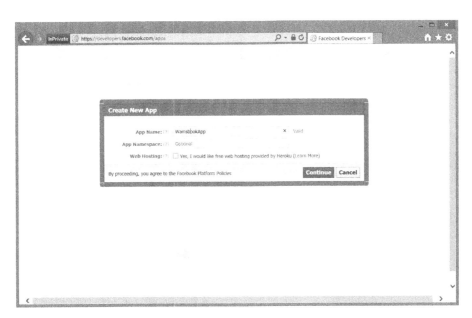

FIGURE 4-5

Enter the name of your application in the first text box. This name must be unique (that is, not used by any other Facebook application), so you might have to get a little creative with the naming. However, the app's name is not visible to anyone who is using your Windows application, so the actual name is not particularly important. Once you have entered the name, click the Continue button. This creates the application and displays your new application's Basic settings form, as shown in Figure 4-6.

Notice that the Site URL property for the Website with Facebook Login region is filled in. This value (the field is revealed when you click the Website with Facebook Login label) is the URL for your mobile service.

Before you click the Save Changes button at the bottom of the screen, make a note the App ID and App Secret that appear at the top of the page. You will need this information shortly. As well, you should treat the App Secret as a critical piece of security information — protect it because it is used as part of the authentication for your application. When you have copied down the App ID and App Secret, click Save Changes to update your Facebook application settings. The next steps for connecting your application to Facebook's authentication mechanism are discussed after the section on Twitter.

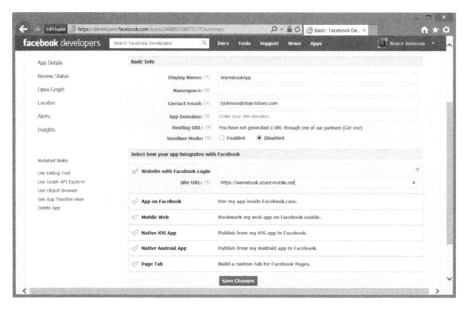

FIGURE 4-6

Google

The process of setting up Google to allow for authentication with WAMS is quite similar to the Facebook process. Before you can begin, you must have a Google account. You can register for a Google account at http://accounts.google.com. Once you have an account, navigate to the URL https://code.google.com/apis/console to start creating your project definition.

Log in with your Google account and you are taken to the page shown in Figure 4-7.

> **NOTE** *If you have already created a project under your Google account, you might be taken to a page that displays information about your projects (similar to Figure 4-11). If this is the case, the button to create a new client ID appears at the bottom of that list. Click that button and the screen shown in Figure 4-9 appears.*

Because the only user interface element on this page is Create Project, it's a safe bet that clicking that is the next step. So, go ahead and do that to bring up the dialog box shown in Figure 4-8.

Authentication Options in WAMS

FIGURE 4-7

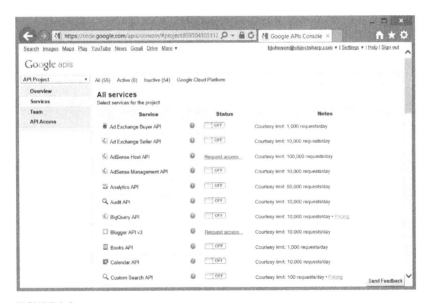

FIGURE 4-8

On the navigation menu that appears on the left of the screen, click the API Access link, and then click the Create an OAuth 2.0 Client ID button that appears in the middle of the screen. At this point, Figure 4-9 appears and the provisioning of your application begins in earnest.

CHAPTER 4 AUTHENTICATION OPTIONS IN WAMS

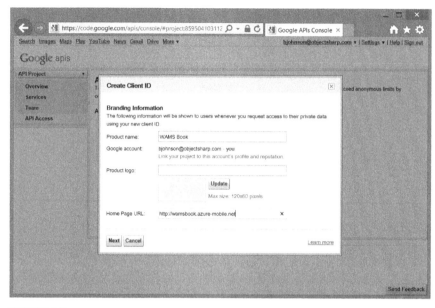

FIGURE 4-9

Provide the name of your project in the appropriate text box. Your Google account is already filled in. Also, if you're feeling in a particular marketing bent, you can provide a logo for your application in this form. Once you have filled in all of the required information, click the Next button in the bottom right to move to Figure 4-10.

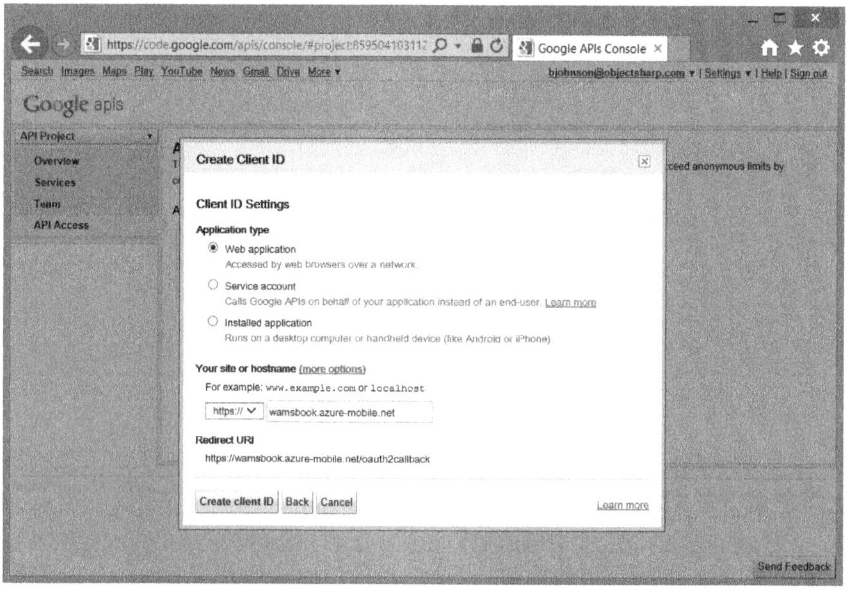

FIGURE 4-10

Authentication Options in WAMS

In the Client ID Settings sections are a number of fields that need attention. To start, set your Application Type to Web Application. This is because, regardless of the client platform, the request for authentication will take place across a network. Set the Your Site or Hostname value to the URL for your mobile service. (The values in Figure 4-10 are for a sample mobile service located at http://wamsbook.azure-mobile.net.) When you have filled in these values, click the Create Client ID button. A summary screen, shown in Figure 4-11, appears.

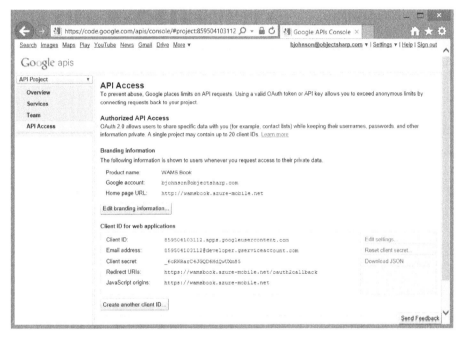

FIGURE 4-11

On the summary screen, you will find both the client ID and the client secret for your application. Make a note of these values because you will need to provide them to WAMS in the steps that are discussed later.

Microsoft Account

The next STS provider is Microsoft. Specifically, you learn how to set up Microsoft Account to authenticate your users. As with the other providers, you need to have a Microsoft Account in order to register your application. Because Microsoft Account is the new name for what used to be known as Microsoft Live ID, using your Microsoft Live ID is fine as well.

Start by going to http://dev.live.com. If you haven't signed in, there is a Sign In link in the top-right corner of the page. And in the menu at the top of the page (after you have signed in), select My Apps to get to Figure 4-12.

FIGURE 4-12

Underneath the My Applications label, there is a small link called Create Application. Click this link to go to the page shown in Figure 4-13.

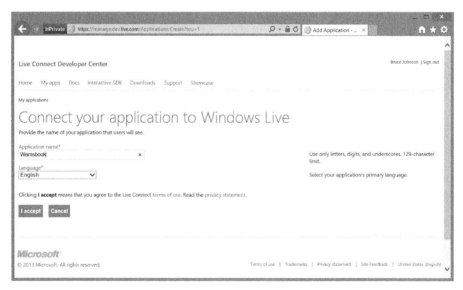

FIGURE 4-13

Authentication Options in WAMS

Provide the name of your application along with the primary language that your application is using. Review the terms and conditions of usage and, if acceptable, click the button labeled I Accept. The page that contains the API Settings for your application appears (Figure 4-14).

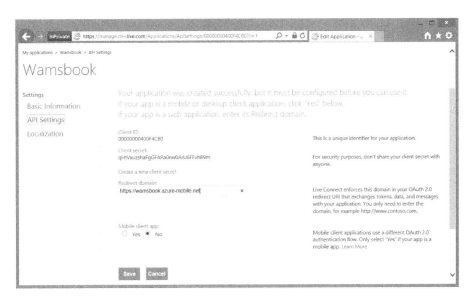

FIGURE 4-14

This page already has the client ID and client secret that you need for the next step. Make sure to make a note of them. But before your application is completely configured, you need to add the redirect domain, which located just below the Client Secret field. Enter the URL to your mobile service, and then click Save to persist the configuration changes.

Twitter

The last of the four STS providers is Twitter. As with the other three, you need to have a Twitter account in order to define your application. So the starting point is to go to http://www.twitter.com and set up an account if you don't already have one.

Navigate to http://dev.twitter.com/apps to start the process. After you have signed in with your credentials, you see the page shown in Figure 4-15.

When you click the Create a New Application button in the top right, the Create an Application page shown in Figure 4-16 appears.

CHAPTER 4 AUTHENTICATION OPTIONS IN WAMS

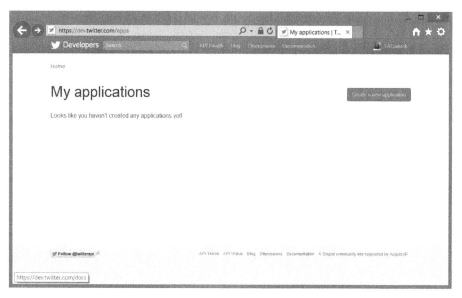

FIGURE 4-15

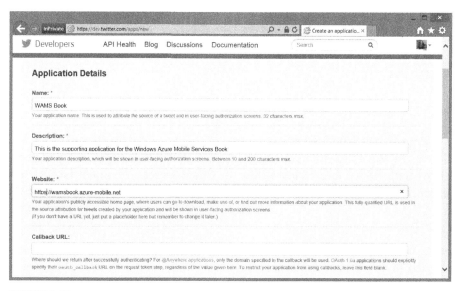

FIGURE 4-16

On this page you provide the details about your application. Fill in the Name and Description fields, and under Website add the URL to your homepage. If you don't have one, the URL to your mobile service will work here as well. The Callback URL is used to call back to your application after login, but it is not required and there is no need to provide one. Scroll down to the bottom of the page to reveal the remaining fields (Figure 4-17).

70

Authentication Options in WAMS

FIGURE 4-17

Read the terms and conditions for use of Twitter as an SRS provider and if you accept them, click the Yes, I Agree check box. Enter the correct characters for the CAPTCHA that is present, and then click the button labeled Create Your Twitter Application. After a few moments, the main page for your application appears (see Figure 4-18).

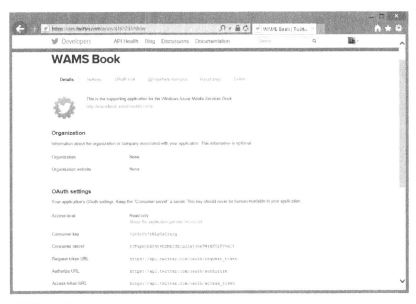

FIGURE 4-18

At the bottom of the page, you will see Customer key and Customer secret values. These are the values you need to use in the next section. However, there is one additional step that needs to be taken. Click on the Settings tab at the top of the page. The settings for your application (shown in Figure 4-19) appear.

FIGURE 4-19

Scroll down to the middle of the page. You will find a check box labeled Allow this application to be used to Sign in with Twitter. Check this and then click on the Update this Twitter Application's Settings button at the bottom of the page.

Configuring Your Service for Authentication

To this point, you have registered your application with one (or more) STS providers. And for each SRS, you have collected a value for your application ID and secret. It's time to add those to the mobile service configuration.

Go through the Azure Management Portal to get to the list of mobile services and then to the starting page for your mobile service. One of the links across the top is Identity. Click it and the page shown in Figure 4-20 appears.

On this page, you will see a section for each of the STS providers. In each section, you will see a space where you can enter the Client ID and Client Secret (or the name that corresponds to the ID and secret for each provider). For each of the providers that you want to support, fill in these values using the information preserved from the registration process. When you have entered all of the details for the desired providers, click the Save button at the bottom. Your mobile service is now ready to handle authentication. Well, almost.

Authentication Options in WAMS

FIGURE 4-20

The last step in the process is to set the security on your table to require that the user must be authenticated to access the table. In Figure 4-21, you see the security page for a given table. To get here, start at the page for your mobile service, click the Data link at the top, select the table that you want to secure, and then click the Permissions link.

FIGURE 4-21

For the operations that you want to secure, make sure that the Only Authenticated Users option has been selected in the combo box. Click the Save button at the bottom when you are finished.

Authentication on the Client Side

If you think back to the flow that is diagrammed in Figure 4-2, you might realize that the beginning of the authentication process must start with the client application. Specifically, the application must present a user interface that lets the users enter their credentials for the STS provider.

By using the mobile services SDK, the mechanism for doing this is straightforward. To start with, add a Login button to the form and wire up a handler for the click event. The XAML will look like the following:

```
<Button Name="btnLogin" Content="Login" Click="btnLogin_Click" />
```

The click event handler is shown in the following code:

C#

```csharp
MobileServiceUser user = null;

private async void btnLogin_Click(object sender, RoutedEventArgs e)
{
   if (user == null)
   {
      try
      {
          user = await App.MobileService.LoginAsync(
             MobileServiceAuthenticationProvider.MicrosoftAccount);
      }
      catch (InvalidOperationException)
      {
          // the authentiation failed
      }
   }
}
```

VB.NET

```vb
Dim user As MobileServiceUser = Nothing

Private Async Sub btnLogin_Click(sender As Object, e As RoutedEventArgs)

   If user Is Nothing Then
      Try
```

Authentication Options in WAMS

```
        user = Await App.MobileService.LoginAsync( _
            MobileServiceAuthenticationProvider.MicrosoftAccount)

    Catch InvalidOperationException

        // the authentiation failed

    End Try
  End If

End Sub
```

You need to be aware of a couple things here. First, the trigger for prompting users for their credentials is the call to the `LoginAsync` method. The appearance of the dialog box will depend on the platform in which the client application is running. Also note the `MobileServiceAuthenticationProvider` enumeration. Figure 4-22 illustrates the dialog box for a Windows Store app using Microsoft Account as the provider. The method will change depending on the provider that you want the value passed into so if you want to utilize different provider, pass in a different value.

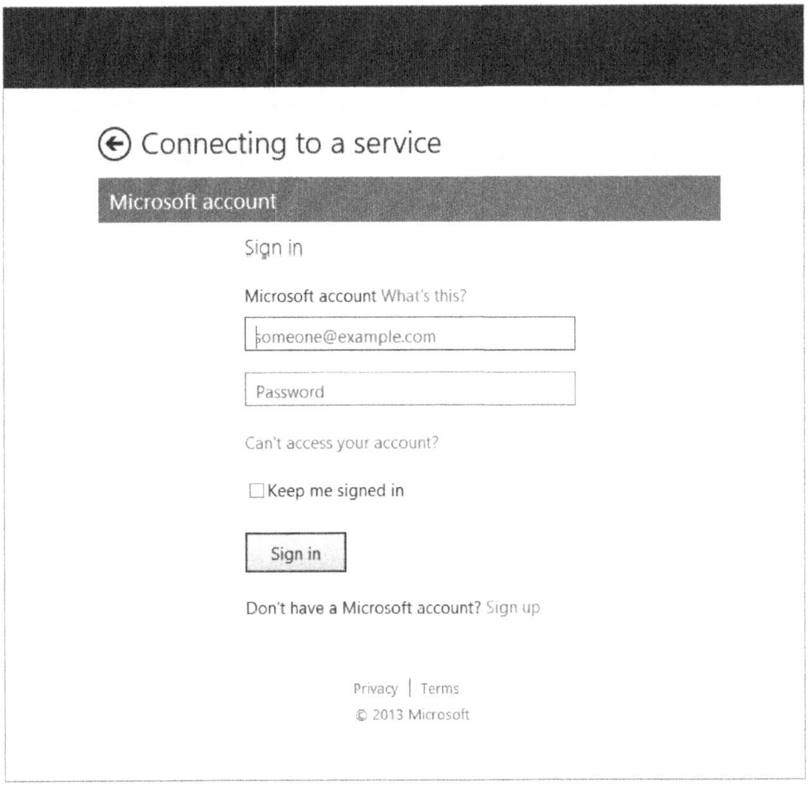

FIGURE 4-22

Another thing to be aware of is the *user* variable, which is defined at the class level. This variable is assigned to the result of the LoginAsync call. This value contains (not surprisingly) the current user if one has logged in. You'll notice that the login dialog box is invoked only if the user is null.

However, it also contains additional information that can be used to persist the logging-in functionality beyond the end of the current application. Specifically, the two properties that you can use to implement this are the UserId and the MobileServiceAuthenticationToken.

The key to making this work is found in a different way to create a MobileServiceUser object. Specifically, you can pass the user ID into the constructor and then assign the MobileServiceAuthenticationToken property directly. This new instance of a MobileServiceUser can then be assigned to the CurrentUser property of the MobileService object. The following code provides an example:

C#

```
MobileServiceUser user;

user = await App.MobileService.LoginAsync(
   MobileServiceAuthenticationProvider.MicrosoftAccount);


MobileServiceUser newUser = new MobileServiceUser(user.UserId);
newUser.MobileServiceAuthenticationToken = user.MobileServiceAuthenticationToken;

App.MobileService.CurrentUser = newUser;
```

VB.NET

```
Dim user As MobileServiceUser

user = Await App.MobileService.LoginAsync( _
   MobileServiceAuthenticationProvider.MicrosoftAccount)


Dim newUser As New MobileServiceUser(user.UserId)
newUser.MobileServiceAuthenticationToken = user.MobileServiceAuthenticationToken

App.MobileService.CurrentUser = newUser;
```

At this point, any communications through the proxy class will utilize the values found in newUser. What this means is that if you take the user ID and the authentication token that is created after a successful login and persist that information locally, you can implement the functional equivalent of "Remember Me." That is to say, that the next time the application is started, you could retrieve the user ID and authentication token, create the MobileClientUser object, and assign it to CurrentUser.

Authentication Options in WAMS

Of course, if you plan on implementing this functionality, it's important to remember that the user ID and authentication token information fall into the category of "sensitive." In other words, make sure that you take the appropriate precautions to ensure that this information isn't unintentionally exposed.

Troubleshooting Authentication

As you go through the process of configuring authentication, you may occasionally run into issues. This is not unexpected because a number of pieces need to be set up correctly for the authentication process to work. If one of those pieces is not set up properly, you see that surface in your application any one of a number of issues.

Debugging these issues is quite challenging if you go through your application. The benefit of the set of proxy classes provided by WAMS is that it's very easy to perform basic (or even decently complex) functionality. However, when something goes wrong, that abstraction can get in the way of identifying the fundamental problem.

The trick is to find a way to get around the abstraction that is implemented in the proxy classes. Fortunately, the authentication flow uses a series of HTTP GETs, which means that you can use a browser to trigger the request and evaluate the response using your favorite web communication debugging tool.

When an authentication request is sent to your mobile service, the endpoint is `https://`*servicename*`.azure-mobile.net/login/`*provider*, where *servicename* is the name of your mobile service and *provider* is the name of the STS provider (that is, `facebook`, `google`, `microsoftaccount`, or `twitter`).

The first response from this endpoint is to redirect the user (using an HTTP 302) to the provider's login page. Once you have logged in, you will receive an HTTP 302 response that redirects back to your mobile service endpoint and contains an authentication token. The mobile service validates the provider's authentication token, generates a token of its own, and sends it back to the client.

So how does this help you troubleshoot? Well, eventually there will be a response back to your browser with a JSON body. That body contains information that may not make it into the `InvalidOperationException`, which is raised if the login attempt is unsuccessful. For example, the following is retrieved if the Google account is not set up properly:

```
{"code":401,"error":"Error: Logging in with google is not enabled."}
```

By reading the error message that is available through the JSON body, you are in a better position to correct the problem. As an aside, you can use any tool you prefer to read the JSON body. Fiddler is a common favorite, as well as the developer tools that are available for Chrome or Firefox browsers. Even Internet Explorer allows you to see the JSON body; it just asks if you want to download the file that contains the JSON body first.

SUMMARY

While authenticating users probably doesn't rise to the level of basic functionality in the way that reading and writing data does, there is no question that a large percentage of mobile services will utilize authentication in one way or another. Using the popular authentication providers as described in this chapter go a long way towards reducing the friction that imposing another set of credentials would create. And it allows for the socialization of your applications functionality by seamlessly incorporating your app into Facebook, Twitter, etc.

In the next chapter, we look at how REST-based requests can be used to perform data access functionality. While the generated proxy client is nice, the availability of REST means that almost every software application, regardless of platform, can utilize WAMS.

5
Using REST to Access WAMS Data

IN THIS CHAPTER:

➤ Understand the basis concepts and verbs used in Representational State Transform

➤ Use REST to perform CRUD functionality against your mobile service

➤ Send administrator authenticated requests to your mobile service

WROX.COM CODE DOWNLOADS FOR THIS CHAPTER

The wrox.com code downloads for this chapter are found at http://www.wrox.com/go/windowsazuremobileservices on the Download Code tab. The code is in the Chapter 5 download and individually named according to the names throughout the chapter.

The ability to use client-side libraries to access your mobile service is certainly convenient. It's nice to be able to write a single line of code that updates and retrieves data. However, in some situations client-side libraries are not available, but the need to access the Windows Azure Mobile Services (WAMS) data remains. This is the gap that REST-based access is designed to fill.

REPRESENTATIONAL STATE TRANSFER

Representational State Transfer (REST) is a relatively recent addition to the toolbox of web development. It is an architectural model with a number of fairly distinctive attributes. Conceptually, REST provides a very clean separation of client interface from the actual implementation on the server. It's a stateless interface, which is quite valuable in the web world. However, the attribute that is most characteristic of REST is the requirement for a uniform interface, which is both an interesting and powerful concept in itself.

To understand why the uniform interface makes such a difference, consider how you might have approached the creation of a web service in the past. It would be quite common to define a number of CRUD methods, such as CreateCustomer or UpdateCustomer. For retrieving information, you might have GetCustomersByName and GetCustomersByRegion methods. Or, if you were particularly ambitious, you might create a more generic GetCustomers method where the conditions for retrieval are passed in as parameters. Of course, you'd have to do the same thing with Orders and Products as well. Or, if the Products web service was created by someone else, perhaps the methods would be called InsertProduct and ModifyProduct.

In other words, when you manually create the methods that are used to perform CRUD operations, the potential for an overly complicated and inconsistent interface is quite high.

The idea of a uniform interface is to remove this potential for complexity. When you're dealing with web applications, the REST interface is defined through the ubiquitous HTTP protocol, which includes the verbs GET, POST, PUT, and DELETE. By using these verbs for the CRUD functions and defining a format for the URI so that you can specify the target object and the identifier, many of the issues surrounding the aforementioned web service are eliminated. When you include the ability to add conditional querying to REST, you end up with a very powerful, yet still uniform, mechanism for exposing common data functionality.

The following sections consider the different verbs and the type of functionality that each one of them offers.

GET

You can use the GET verb in two different ways, depending on the target. If the target is the name of a collection, a list of the URIs and optionally other information about the members in the collection are retrieved. If the target is an individual item, information about that specific item is retrieved. The difference between these two targets is best illustrated through examples. Consider the following URIs:

```
http://wamsbook.azure-mobile.net/tables/gameResult
http://wamsbook.azure-mobile.net/tables/gameResult/1
```

In the first case, the target is the collection with the name `gameResult`. The result from passing that URI to the service would be a list of the URIs in the `gameResult` collection. The second URI would retrieve the information about the URI with the identifier of 1.

POST

Again, you can interpret the purpose of the POST message in two ways, but the second is rarely used. To be able to craft POST requests, it's important to be aware of how the POST message is structured.

In the case of the GET, all of the information to retrieve the appropriate object is supposed to be in the URI. With the POST, the information is in both the URI and the body of the request.

The POST message is used to create a new entry in a collection. Consider the following URI:

```
http://wamsbook.azure-mobile.net/tables/gameResult
```

When sent as a POST with no body, this request creates a new, empty `gameResult` object and returns the ID for the new record. If the body of the POST request contains a `gameResult` object, it creates a new `gameResult` using the values in that object. The identifier for the new object will be returned.

The second type of POST message involves specifying an identifier in the URI. Technically, the outcome of processing the message is to treat the identified object as if it were a collection and create a new object within that collection. If you think about it, this would make the `gameResult` collection really a collection of collections. Although this approach is definitely consistent, it is not typically how one thinks about the persistence and access of objects through the REST (or any other) mechanism.

PUT

The PUT message is conceptually the equivalent of an update. Again, you can interpret it two ways depending on the URI. If the URI targets the collection as a whole, the implementation replaces the specified collection with the collection that is included in the body of the message.

If, on the other hand, the URI includes the identifier, an object is either modified (if the identified object already exists) or created (if it does not). The values used in the updated object are taken from the values found in the body of the message.

DELETE

The DELETE message is probably the simplest of the verbs to explain: The processing of the message results in the deletion of either the entire collection or the specified object. Naturally, the function performed depends on whether the URI specifies an individual object or the entire collection, but the result is pretty much what you would expect.

For completeness, the description of the fundamentals of REST should consider a couple of additional concepts. First, by definition, the PUT and DELETE verbs must be *idempotent*. A verb is considered idempotent if multiple executions of the same request result in the same state in the underlying store and return the same result. So, sending the DELETE message twice will not cause the data store to "look" different after the second message than it did after the first. Sending a PUT message twice has the same effect.

Additionally, the GET verb must be *nullipotent*. The concept here is that there cannot be any side effects from executing the GET command. Nothing in the data store should be changed, and the results from consecutive GETs will be identical.

Finally, just so you can get your head around the URI/body relationship in some of the messages, consider the following snippet.

This is the message that would be sent to update the `gameResult` object with an identifier of 100. You can see that the first line has the verb and URI:

```
PUT https://wamsbook.azure-mobile.net/tables/gameresult/100
Accept: application/json
Content-Type: application/json
Host: wamsbook.azure-mobile.net
Content-Length: 62

{"id":100,"score":153,"resultDate":"2013-02-16T05:57:21.242Z"}
```

Then, after the various headers, the body of the message contains the serialized version of the new object.

As you think about these messages and consider the need to update a particular object, you might stumble on a potential problem (not a problem, so much as a potential inefficiency). From the preceding example, it appears that to update an object, you need to send the entire object in the body of the message. For something the size of gameResult, that's not such a big deal. But if you had a Customer object or a BillOfMaterials object, the size of the message could get significant. If you're modifying only a single field in the object, that seems like an overly heavy payload for the purpose.

As a result, there is another verb that is supported in most (but not all) REST services, and most importantly for the topic of this book, it's supported by WAMS. That verb is PATCH. Functionally, the PATCH verb acts in a similar manner to the PUT message in that it is used to update an existing object. If the object specified in the URI doesn't exist, an error response is received. But if the object is there, then instead of expecting to have the complete object in the body, it updates only those properties that are present. If all that was required was to update object 100 in the gameResult collection, the message would look like the following:

```
PATCH https://wamsbook.azure-mobile.net/tables/gameResult/100
Accept: application/json
Content-Type: application/json
Host: wamsbook.azure-mobile.net
Content-Length: 13

{"score":175}
```

REST AND WAMS

You've gone over the basics of REST, so now it's time to map what has been covered in general onto the specifics of WAMS. The simplest way to do this is to compare the statements used to perform CRUD functionality. Windows Azure Mobile Services supports the request formats shown in Table 5-1.

TABLE 5-1: REST Operations for WAMS

REQUEST	FUNCTIONALITY
GET https://*servicename*.azure-mobile.net/tables/*table*/	Retrieves all of the records in the table
GET https://*servicename*.azure-mobile.net/tables/*table*/1	Retrieves a single record from the table, as specified by the identifier

`GET https://servicename.azure-mobile` `.net/tables/table?$filter`	Retrieves one or more records from the table using an OData-formatted filter
`POST https://servicename.azure-mobile` `.net/tables/table/`	Adds a new record to the table
`PATCH https://servicename.azure-mobile` `.net/tables/table/1`	Modifies an existing record in the table
`DELETE https://servicename.azure-mobile` `.net/tables/table/1`	Deletes a record from the table

These request messages align closely to the messages that were covered earlier in the chapter. The biggest discrepancy is the use of the PATCH verb instead of PUT to handle all of the updates. To demonstrate how these requests might be used in practice, the following section considers some code.

Getting Data

The technique used to construct a REST request depends greatly on the platform. For .NET, the `HTTPWebRequest` and `HTTPWebResponse` classes are the most convenient choices. If you are writing in PHP, Objective-C, or any other environment, classes are available that provide corresponding functionality. For JavaScript, there is a challenge that is covered at the end of this section. The .NET code for performing a GET is the following:

C#

```
WebClient request = new WebClient();

request.Headers.Add("Accept", "application/json");
var response = request.DownloadString(
   new Uri("https://wamsbook.azure-mobile.net/tables/gameresult"));
```

VB.NET

```
Dim request As New WebClient()

request.Headers.Add("Accept", "application/json")
Dim response = request.DownloadString( _
   new Uri("https://wamsbook.azure-mobile.net/tables/gameresult"))
```

If you have written any .NET code to initiate an HTTP request, this code will be moderately familiar. If you have used the `HttpWebRequest`/`HttpWebResponse` classes, the flow is the same, although the details have changed.

The `WebClient` object is used as the starting point for making the request. The `Accept` header on the request is set to `application/json` to indicate that the application would like to receive a response in JSON. Then the `DownloadString` method is called, passing in the URI that is the target of the request. The result of `DownloadString` is that an HTTP GET message is sent to the target, and the response (which for WAMS will be a JSON object) is converted to a string. So the question becomes "how do you get from that string back into an object that you can use?" Unfortunately, it's not easy to arrive at the straightforward answer.

Ultimately, the answer to the question is to use a third-party library called Json.NET to perform the conversion. So that you're able to explain to your boss why this third-party component is useful, let's walk through the process.

Typically, in .NET, you have two mechanisms for deserializing a JSON object: a `JavaScriptSerializer` class and a `DataContractJsonSerializer` class. These classes perform similar functions, in that they are both, more or less, capable of converting a .NET object into its JSON representation and then converting it back again. However, it's the "more or less" clause that has the potential to get you into trouble.

The issue relates to how data types are handled during the conversion — or more specifically, with how dates are handled during the conversion.

When you pass a date to a mobile service using one of the proxy classes discussed in the earlier chapters in this book, that date is passed along as an ISO-formatted date. If you look in the data store for a table that contains such a date (and the `gameResult` table that you have been working with does have a date field), you will see that the values are UTC-formatted dates. This is what you expect to happen when you use a date.

However, the two serializers that are native to .NET don't readily recognize UTC-formatted dates. Instead, they like to see dates in the form of `Date(12312331323)`. Therefore, when the WAMS REST API returns the UTC-formatted dates, a serialization exception is thrown.

The solution to this problem is to utilize the previously mentioned third-party library. Although others might work, the Json.NET library from James Newton-King is both fast and flexible. It has the benefit of being part of a NuGet package, making it very easy to include in whatever application you are running. From Visual Studio 2012, right-click your project and select the Manage NuGet Packages option, as shown in Figure 5-1.

In this dialog box, enter **json.net** into the Search Online box in the top-right corner. Select the Json.NET item that appears and click the Install button. This adds the elements necessary for Json.NET to your project.

Now that Json.NET is available, you can quickly put it to use. The following code extends the previous code snippet to deserialize the objects into instances of a .NET class:

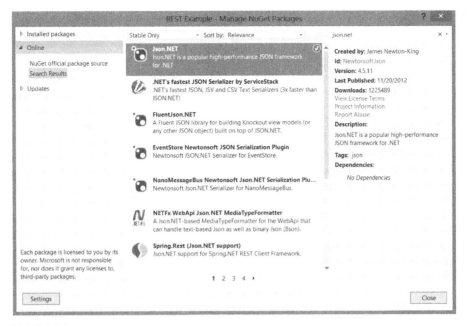

FIGURE 5-1

C#

```
WebClient request = new WebClient();

request.Headers.Add("Accept", "application/json");
var response = request.DownloadString(
    new Uri("https://wamsbook.azure-mobile.net/tables/gameresult"));
var scores = JsonConvert.DeserializeObject<List<gameResult>>(response);
```

VB.NET

```
Dim request As New WebClient()

request.Headers.Add("Accept", "application/json")
Dim response = request.DownloadString( _
    new Uri("https://wamsbook.azure-mobile.net/tables/gameresult"))
Dim scores As JsonConvert.DeserializeObject(Of List(Of gameResult))(response)
```

One final word of caution regarding the use of the REST API and Json.NET: Json.NET does *not* recognize the use of the `DataMember` attribute to change the name of your property to one that is used by your mobile service. So if you used the `DataMember` attributed class from Chapter 2, you will find that some of the values are not being deserialized properly. On the other hand, Json.NET does not seem to have an issue with case-sensitive property names. In other words, it will deserialize the JSON element of `id` into either the `id` or `Id` property with no need for attributes.

Filtering on GET

The previous section covered the mechanics that get all of the records from the table. Based on the information in Table 5-1, retrieving an individual record is just a matter of adding the identifier for that record onto the URL. But what if you want to retrieve some, but not all, of the records? Or sort the results prior to retrieval? Fortunately, the WAMS REST API provides a mechanism that enables you to do this and that mechanism utilizes the OData protocol.

The *Open Data Protocol* (OData) specifies a number of recommendations for constructing URIs to identify data, as well as specific query string operators to retrieve the data. It is these query string operators that are used in the filtering process. Table 5-2 contains the query string operators that are supported by WAMS.

TABLE 5-2: REST API Query String Parameters

QUERY STRING PARAMETER	DESCRIPTION
$filter	Indicates the filter predicate that is used to restrict the returned records.
$inlinecount	Returns the number of items that can be returned when the request is processed.
$orderby	Indicates the columns that should be used to sort the records prior to returning them.
$select	Specifies the columns that are to be returned by this request.
$skip	Skips over the specified number of records before returning the remainder of the matching records. This parameter would typically be used in a paging mechanism.
$top	Specifies the number of records to include in the result. This parameter is also used in a paging mechanism.

Because we're talking about filtering, focus on the $filter query string parameter. The format of the filter is as follows:

```
.../tables/gameResult?$filter=score gt 150
```

As mentioned in Table 5-2, $filter is the query string parameter used to specify the conditions that need to be satisfied in order to include a record in the result. The value of the parameter is a predicate expression that is evaluated for each record. A number of operations are supported in this expression, including the major Boolean operations (equal, not equal, greater than, less than, and so on). You can specify multiple conditions using and, or, and not, along with parentheses to group the conditions as required:

```
.../tables/gameResult?$filter=(score gt 150) or
   (gameDate gt datetime'2013-03-16T00:00:00Z')
```

This example also demonstrates how the most complex of the intrinsic data types are handled. As you can see, numeric values are specified with no delimiter. Strings, although they don't appear in the example, use a single quote (') to denote the boundaries. Dates, however, use the notation found in the code snippet, which (at a functional level) converts the string value representation of a date into a `DateTime` object.

Many more options and functions are available to use in the WAMS REST API for retrieval. For a list of the options that are recommended by the OData specification, visit http://www.odata.org/developers/protocols/uri-conventions.

Updating Data

The process of updating data using the WAMS REST API has fewer options than GET, in terms of the mechanics. Typically, you have to deal with two variables. The first is how to specify the verb that is used to submit the request. Fortunately, that's straightforward. When using the `WebClient` object you have a number of `Upload` methods, including `UploadData`, `UploadString`, and `UploadFile`. All of these methods have a signature where one of the parameters is the HTTP verb. Because it fits well with Json.NET, the examples here use `UploadString`.

The second variable relates to the creation of the body for those methods (POST and PATCH) that require a JSON object. As it turns out, you have already discovered a tool that will help greatly in this process — namely, the Json.NET library.

As mentioned earlier in this chapter, the POST and PATCH methods take a JSON object as the body. The property names and values in the object are the properties and values that will be inserted or updated in your mobile service, so the key is to be able to serialize the object that is to be updated. The following code provides an example:

C#
```
WebClient request = new WebClient();

request.Headers.Add("Accept", "application/json");
gameResult gameResult = new gameResult()
    {Id=5, Score = 221, GameDate = DateTime.Now };
string body = JsonConvert.SerializeObject(gameResult);

var response = request.UploadString(
    "https://wamsbook.azure-mobile.net/tables/gameResult", "POST", body);
var newGameResult = JsonConvert.DeserializeObject<gameResult>(response);
```

VB.NET
```
Dim request As New WebClient()

request.Headers.Add("Accept", "application/json")

Dim gameResult = New gameResult() With _
    { .ID=5, .Score=221, .ResultDate = DateTime.Now }
```

```
string body = JsonConvert.SerializeObject(gameResult)

var response = request.UploadString(
   "https://wamsbook.azure-mobile.net/tables/gameResult", "POST", body)
Dim scores As JsonConvert.DeserializeObject(Of gameResult)(response)
```

The `Serialize` method is used to convert the managed code object into its JSON representation. The result is passed as one of the parameters into `UploadString`. Now when the `UploadString` method is invoked, the request, the verb, and the body, is sent to your mobile service, where it is processed appropriately.

Authentication in REST

To this point, the REST requests have presumed that you have access to all of the CRUD functionality in the mobile service. If you map this back to the permissions that are applied to the tables in your mobile service, you have presumed `Everyone` is the value. However, there will be times when anonymous access is not enabled and there is still a need to use REST. This section describes how that can be accomplished through the REST API.

To get started, consider the different levels of authentication and how the necessary information is conveyed to WAMS. If you recall, there are four levels of authorization: Everyone, Anybody with the application key, Only Authenticated Users, and Only Scripts and Admins. To this point, because you have provided no additional details, the presumption was that `Everyone` was allowed to make the request. To determine the other levels, you need to provide an additional header in the HTTP request.

Application Key

The application key is the key that uniquely identifies your WAMS mobile service. When the client-side code was covered in Chapter 2, the application code was part of the constructor for the `MobileServiceClient` object. The following code shows that constructor:

C#

```
public static MobileServiceClient MobileService = new MobileServiceClient(
          "https://wamsbook.azure-mobile.net/",
          "FGNIFOZqblbzCMsCYRpLIHVthJxala84"
          );
```

VB.NET

```
Public Shared MobileService As MobileServiceClient = _
   New MobileServiceClient( _
      "https://wamsbook.azure-mobile.net/", _
      "FGNIFOZqblbzCMsCYRpLIHVthJxala84" _
   )
```

The application key is the ugly string that starts with "FGNIF". To pass this key along with the REST call, it is placed into a header with the name X-ZUMO-APPLICATION. The following code demonstrates how it would be included in a request:

C#

```
WebClient request = new WebClient();

request.Headers.Add("X-ZUMO-APPLICATION",
   "FGNIFOZlbzCMsCYRpLIHVthJxala84");
request.Headers.Add("Accept", "application/json");
var response = request.DownloadString(
   new Uri("https://wamsbook.azure-mobile.net/tables/gameResult"));
var scores = JsonConvert.DeserializeObject<List<gameResult>>(response);
```

VB.NET

```
Dim request As New WebClient()

request.Headers.Add("X-ZUMO-APPLICATION", _
   "FGNIFOZqbYRpLIHVlbzCMsCthJxala84")
request.Headers.Add("Accept", "application/json")
Dim response = request.DownloadString( _
   new Uri("https://wamsbook.azure-mobile.net/tables/gameResult"))
Dim scores As JsonConvert.DeserializeObject(Of List(Of gameResult))(response)
```

In case you are curious as to why the header includes the term ZUMO, it's not as mysterious as you might hope. ZUMO was the code name for Windows Azure Mobile Services while it was under development.

Authenticated Users

The mechanism for providing the information necessary for authenticated users is, more or less, the same as for the application. A header needs to be provided along with the request. The differences are that the name of the header is X-ZUMO-AUTH, and the value of this header is an authentication token. The real trick with this level is how the authentication token is determined.

To start, remember that users become authenticated by providing their credentials to a third-party authentication provider, such as Facebook, Google, Microsoft Account, and Twitter. However, the authentication token that is provided to you by these providers needs to be associated with a user, and this association needs to be done by WAMS. To get to this point, the authentication token needs to be sent to WAMS. WAMS then sends the authentication token to the third-party provider and gets back the identity of the user. This flow of information is initiated by sending a particular request to the REST API.

To start with, you need to get the authentication token for the third-party provider, so your application needs to log in to one of these providers on its own. This is different from the client-side login mechanism, which automatically displays the page used to collect the user credentials. As part of the response to that login process, you will receive an authentication token.

To get your mobile service to recognize the token and, more importantly, to map it to a user ID, you make a REST request. The URI that is the target of the request is `https://servicename.azure-mobile.net/login?mode=authenticationToken`, where `servicename` is the name of your mobile service. There is a query string parameter named `mode` on this URI, which is used to specify how the authentication information is to be returned. At this version of WAMS, the only supported mode is `authenticationToken`.

This login request uses a POST verb and there is a body for the request. Specifically, the body contains a JSON object that has the value of the authentication token. The format of the body is as follows:

```
{"authenticationToken":"eyJhbGciOiJIU..."}
```

Putting all of the pieces together, the following code demonstrates how to make this call:

C#

```csharp
WebClient request = new WebClient();

request.Headers.Add("Accept", "application/json");
var response = request.UploadString("https://wamsbook.azure-mobile.net/login",
    "POST", "{\"authenticationToken\":\"eyJhbGciOiJIU...\"}"));
```

VB.NET

```vbnet
Dim request As New WebClient()

request.Headers.Add("Accept", "application/json")

Dim response = request.UploadString("https://wamsbook.azure-mobile.net/login", _
    "POST", "{\"authenticationToken\":\"eyJhbGciOiJIU...\"}"))
```

The response to this request, presuming it is successful, includes both the original authentication token, along with the mapped user ID. However, in subsequent requests, all that is required is for you to provide the authentication token in the `X-ZUMO-AUTH` header, as shown in the following code:

C#

```csharp
WebClient request = new WebClient();

request.Headers.Add("X-ZUMO-AUTH",
    "eyJhbGciOiJIU...");
request.Headers.Add("Accept", "application/json");
var response = request.DownloadString(
```

```
    new Uri("https://wamsbook.azure-mobile.net/tables/gameResult"));

var scores = JsonConvert.DeserializeObject<List<gameResult>>(response);
```

VB.NET

```
Dim request As New WebClient()

request.Headers.Add("X-ZUMO-AUTH", _
   "FGNIFOZqbYRpLIHVlbzCMsCthJxala84")
request.Headers.Add("Accept", "application/json")
Dim response = request.DownloadString( _
   new Uri("https://wamsbook.azure-mobile.net/tables/gameResult"))
Dim scores As JsonConvert.DeserializeObject(Of List(Of gameResult))(response)
```

Administrative Access

The fourth level that was mentioned earlier is for Scripts and Admins. To gain administrative access to your mobile service, you must provide a master key. The mechanism for providing that key should be familiar to you by now. The key is passed into your mobile service through a header that is included with the request. The name of the header is X-ZUMO-MASTER and the value is the master key for your application.

The master key is defined across your entire mobile service. It is generated when you first set up the service. To see what it is for your application, go into the Azure Management Portal, navigate to your mobile service, and select your mobile service from the list. At the bottom of the screen, click the Manage Keys link to reveal the screen shown in Figure 5-2.

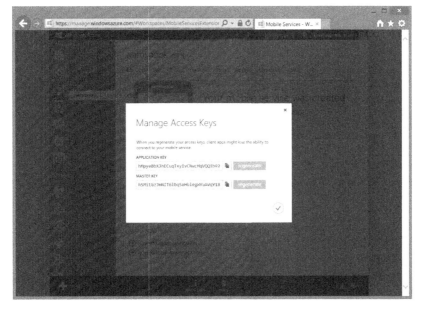

FIGURE 5-2

The master key value is used in the request call, as demonstrated in the following example:

C#
```
WebClient request = new WebClient();

request.Headers.Add("X-ZUMO-MASTER",
    "hMpyeBbXJhECuqTxyivCNwcMqVQQib92");
request.Headers.Add("Accept", "application/json");
var response = request.DownloadString(
    new Uri("https://wamsbook.azure-mobile.net/tables/gameResult"));
var scores = JsonConvert.DeserializeObject<List<gameResult>>(response);
```

VB.NET
```
Dim request As New WebClient()

request.Headers.Add("X-ZUMO-MASTER", _
    "hMpyeBbXJhECuqTxyivCNwcMqVQQib92")
request.Headers.Add("Accept", "application/json")

Dim response = request.DownloadString( _
    new Uri("https://wamsbook.azure-mobile.net/tables/gameResult"))

Dim scores As JsonConvert.DeserializeObject(Of List(Of gameResult))(response)
```

There is one final thing to note about the use of the master key. If you think about the preceding example, you might realize that, with the master key, there was no need to provide any other form of authentication. And, regardless of the security level, the administrator has the rights to add, delete, and change any information that it wants. As a result, make sure that you keep the master key information secure. Do not embed it in the applications that you deliver to clients, for example. Do not transmit it over a channel unless it is secure. In other words, treat it like it's the critical piece of secure information that it is.

SUMMARY

By allowing REST-based requests to be processed by your mobile service, WAMS increases the number of possible clients significantly. While the proxy class is convenient, HTTP and REST and well understood across many technologies. As a result, the universe of potentials clients is quite large.

After expanding the client base, the next chapter moves in the opposite direction. Push notifications depend on an infrastructure that goes beyond both WAMS and your application. As a result, there are limits the clients that can support it. However, when it comes to making your applications come alive, push technologies are both a useful and commonly used platform for doing so.

6
Push Notifications

IN THIS CHAPTER:

➤ Learn the different types of notifications that are supported by WAMS

➤ Register your application with the different notification services

➤ Implement push notification for your application

WROX.COM CODE DOWNLOADS FOR THIS CHAPTER

The wrox.com code downloads for this chapter are found at http://www.wrox.com/go/windowsazuremobileservices on the Download Code tab. The code is in the Chapter 6 download and individually named according to the names throughout the chapter.

One of the fundamental principles of designing applications for either Windows 8 or Windows Phone is the notion of "aliveness." For these applications, this means that there is potentially a constant stream of activity in the Start screen, the Lock screen, or through toast notifications.

To implement this stream, there is a need for periodic updates, and when you architect for periodic updates, you can take two basic approaches. The first is to initiate the updates from the client side using a *pull* or *polling* request, in which the client application makes a request to the service to see if any updates need to be processed. Situations certainly exist for which polling is the appropriate approach, but when you have applications that might not be active, or are operating across a connection that is either slow or costly (or both), polling is not the ideal solution.

The second approach takes the opposite tack. Instead of pulling the information on a regular basis, the data is "pushed" to the application on an as-needed basis. The result is that updates can take place with greater frequency and urgency. This is not to suggest that the updates are in real time or near real time, but with the polling approach, the average age of the notification would be half of the update interval. With *push notifications*, the average age of the notification is just a few seconds.

With this in mind, be aware that Windows will throttle the number of push notifications that are processed by a device, especially if the device is running on battery power or the traffic becomes excessive. For this reason, you shouldn't architect your application such that it requires the guaranteed delivery of push notifications for proper functioning. However, if you can work within that limitation, you'll find that a lot of benefits can be derived from the use of push notifications — not to mention that your application will come alive with activity, and that activity can go a long way toward ensuring user satisfaction.

In the Windows world, push notifications are enabled through *Windows Notification Services* (WNS), working in conjunction with both the sending service and receiving applications. If you are using WAMS with either iOS or Android devices, you will be using either *Apple Push Notification Services* (APNS) or *Google Cloud Messaging* (GCM), respectively. However, the basic flow between the different components, as shown in Figure 6-1, still applies.

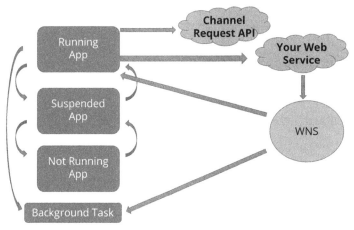

FIGURE 6-1

The application portion of this process is on the left. This is a representation of the different states that an application can be in, from currently running to sort-of running (suspended) to not running at all. Regardless of the state of the application, it can have a background task running as well.

The cooperative portion of the process is related to the top of the diagram. For an application to participate in receiving push notifications, as they relate to WAMS, it needs to perform a couple of steps. The first is to make a request to a web-based API to get a channel. This channel will ultimately be used by WNS to send the push notification to the application, but unless the channel is both created and registered, the service side of the process cannot know who to send the notification to.

After the channel has been retrieved, a request to register the channel is sent to the service side of the process. The service is responsible for detecting when notifications should be sent and then actually performing the transmission. The bulk of the work for this service is in determining what the criteria are for sending notifications, but any coding for this is well beyond the scope of this book.

When the notification is sent along a channel, two components of the applications are able to respond. If the application is running, it receives the notification. If a background task is running, it, too, is capable of receiving the notification. What the application does with the notification information is completely up to the application. But for the moment, it's sufficient that you are able to get notifications from the service to the client on an as-needed basis.

Before getting into the mechanics of pushing notifications, here are the two different types of notifications that are supported by WNS:

- **XML Notifications** — These contain tile updates, badge updates, or a toast notification payload. It's possible for Windows to directly process this type of push notification. What this means is that the update to the various parts of the operating system that were mentioned as "alive with activity" at the beginning of the chapter will happen with little or no effort by the app. It is possible for the app to handle these notifications as well, but the ease in getting the updates to the users makes XML notifications impressive.

- **Raw (or Binary) Notifications** — These contain whatever data the service wants to send. Some limits need to be observed; for example. there is a 5 KB size limit and the data needs to be base64 encoded, but the payload has no limitation of the content. However, because of this flexibility, there is no possibility for Windows to handle the notifications by itself. It is up to the application (or its background task) to process the notifications.

With these preliminaries out of the way, the following section walks through the steps necessary to get your application into the world of push notifications.

REGISTERING YOUR APPLICATION

The first step is to register your application. The details will depend on whether you are using Windows (WNS), Apple (APNS), or Google (GCM) to deliver the notifications. In each case, the delivery platform needs to be aware of your application, so you need to create a unique identifier and a secret and register your application with the platform.

Windows Notification Services

To utilize WNS, you need to register your application with the Developer Center. To start, navigate to the app development page in the Developer Center at http://msdn.microsoft.com/en-us/windows/apps.

A word of warning about the Developer Center: For you to register your application with the Developer Center, you need to have a Developer Account. There are different types of developer accounts. For instance, one can be used to create and publish Windows Store applications. Another allows you to publish Windows Phone applications. But one of these developer accounts is required. There is a cost associated with creating a developer account, although if you have an MSDN subscription, one of the benefits is that you can request a free developer account.

Once you have signed in at the Developer Center using your Microsoft Account (the one associated with your developer account), click the Dashboard link on the top menu. In the Dashboard, click Submit an App on the left navigation menu. This action causes the Submit an App page (Figure 6-2) to appear.

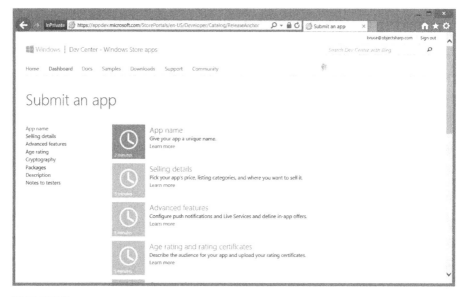

FIGURE 6-2

Before you can generate the required information (the application identifier and secret), you have to give your application a name. Click the App Name icon, fill in the App Name text box with your application's name, and click the Reserve App Name button. Next, click the Advanced Features link in the navigation menu on the left of the screen. One of the advanced features is Push Notifications and Live Connect Services. In that section, there is a link that reads (coincidentally) Push Notifications and Live Connect Services Info. Clicking that link brings up the Push Notifications and Live Connect Services Info screen, as shown in Figure 6-3.

Click the Identifying Your App link and then click the Authenticating Your Service link at the bottom of the page. After following all of these steps, you should be at a page similar to the one shown in Figure 6-4.

Registering Your Application

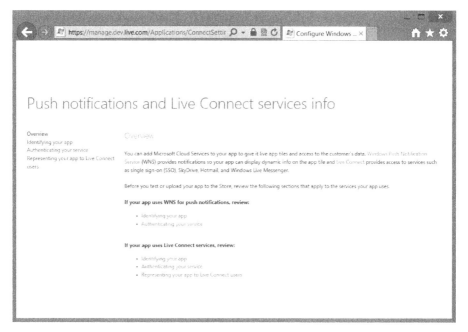

FIGURE 6-3

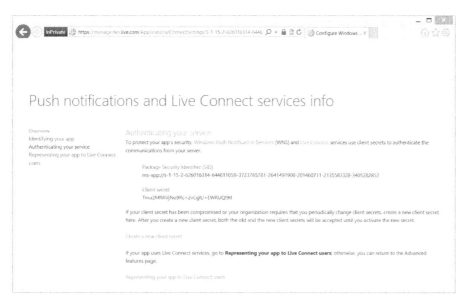

FIGURE 6-4

On this page you can see the Package Security Identifier (SID), along with the Client Secret. Make note of these values for use in the next step.

Google Cloud Messaging

The process for creating the SID and secret for Google Cloud Messaging is very similar. To start, navigate to the Google APIs site at http://code.google.com/apis/console. You will need to log in with your Google ID before the site is visible.

If you have already created a project, you will be directed to the Dashboard page. If not, you will need to create a project. You can find the steps to do this in Chapter 4 in the section titled "Setting Up the Authentication Providers."

Click the Services link in the left-side navigation menu. Scroll down to Google Cloud Messaging for Android and toggle the service on if it has not already been done. You will be prompted to accept a couple of terms of service agreements.

Click the API Access link in the left-side navigation. This takes you to your API Access page, shown in Figure 6-5.

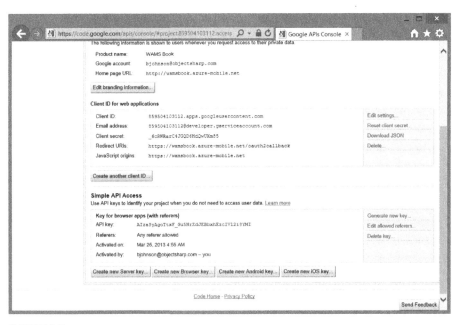

FIGURE 6-5

Click the Create New Server Key button on the bottom of the page. The Configure Server Key for API Project dialog box appears. Click the Create button and the screen shown in Figure 6-6 appears.

Registering Your Application

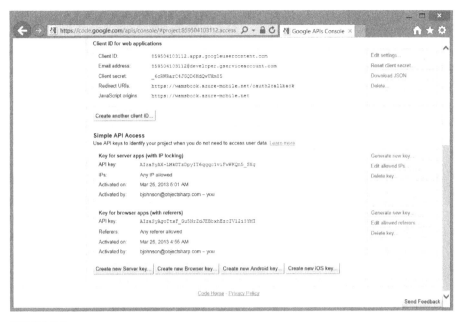

FIGURE 6-6

In the Simple API Access section, there is now an area with the title Key for Server Apps (with IP Locking). Make note of the API Key value found in that section for use in the next step.

Apple Push Notification Services

The starting point for APNS is a little different than GCM or WNS. To provision your application to accept push notifications, you need to generate a Certificate Signing Request (CSR) file. This file will eventually be used in the registration process to generate a signed certificate.

In the Windows world, the mechanism for creating a CSR depends greatly on your development environment. If the machine on which you are developing your application has IIS (either a full version or IIS Express), you can create the certificate request by using Internet Information Services Manager. But, however you do it, you must have a CSR file before continuing to the next step.

In addition, access to the iOS Provisioning Portal is available only to members of the iOS Developer Program, which (at the time of writing) costs $99 per year. You can find the steps to enable APNS and upload your CSR at http://www.windowsazure.com/en-us/develop/mobile/tutorials/get-started-with-push-ios.

CONFIGURING YOUR MOBILE SERVICE

Now that you have all of the information necessary to configure your mobile service so that push notifications can be enabled, you can move on to the next step.

Start by logging in to the Windows Azure Management Portal, and then navigate through the Mobile Services to your mobile service. In the menu at the top, click the Push link. This takes you to the page shown in Figure 6-7.

FIGURE 6-7

The information you enter on this page depends on the service that you plan on using to deliver the notifications. If you use WNS, provide values for the Client Secret and Package SID fields. If you use APNS, click the Upload button and upload the certificate file that was created during the provisioning process. For GCM, fill in the API Key field. When you have finished with the desired fields, click the Save button to save the configuration details and your mobile service is good to go.

Just for completeness, there is no reason that the same mobile service cannot use both WNS and APNS. The mechanism for actually sending a message to a client is different for each of the networks. Which is to say that they use different object. However, both objects can be used in the same script. This would result in notifications being sent to both of the networks, to be broadcast to the appropriate clients, as you would expect.

THE MECHANICS OF PUSH NOTIFICATIONS

Now that your mobile service has been configured, it's time to actually look at the code that is required to perform a push notification. The next sections cover the different stages as outlined at the beginning of this chapter, from creating a channel to pushing out the notification.

Requesting a Channel

Before an application can receive notifications, it must request a channel URI. It is through this URI that the notifications will be sent. Two methods are available to make this request, both of which are in the `PushNotificationChannelManager` class.

The `createPushNotificationChannelForApplicationAsync` method creates a channel URI that is associated with the application's primary tile. Also, any notifications related to toasts, updates, badges, or Raw notifications are automatically directed to the applications main or default tile.

The `createPushNotificationChannelForSecondaryTileAsync` method also creates a channel URI. The difference is that notifications are targeted to a specific secondary tile. The tile that receives these notifications is indicated through a parameter to the method.

Both of these methods return a `PushNotificationChannel` object, which is the application's connection to notifications. A `URI` property uniquely identifies the channel, and an `ExpirationTime` property indicates when the channel will expire.

However, the critical attribute on this object is the `PushNotificationReceived` event. This event is raised every time a push notification is received through this channel, that whether your application supports XML or Raw notifications, you need to handle this event appropriately.

Once the channel has been created, the URI needs to be sent to the service. If you'll recall, it is through this URI that notifications are sent, and when you are using WAMS, this is a relatively easy process. The URI can be stored in a table and then accessed by the server-side script to send out notifications. However, although that is simple, it's not particularly realistic.

So instead, consider the following scenario. You have created an awesome game that combines the playability of Angry Birds with the addictiveness of FarmVille, and as part of that game, you track high scores. Because you believe in making your game as viral as possible, you are utilizing the "friend" functionality found in Facebook and Twitter to track others who are using the game. If a person sets a high score, you want to send a push notification to all of his friends — a "boast," so to speak.

Now think about the information that you need to implement this scenario. Your application makes a channel request. Then the URI for the channel is registered with your mobile service, along with additional information necessary for the service to get the list of friends. For the purpose of this example, the user ID for the authentication user should be sufficient.

One possible approach would be the following: An application registers the channel URI, along with the user ID for the user. The application creates a registration by adding a record to a WAMS table that contains both the URI and the user. In the server-side script, on the insertion or modification of the high score table, get a list of the friends of the person who just set a high score. Using that list of friends, see which of them have registered a URI with your game. If so, use that URI to send the boast notification.

The "registration" of the URI along with the user ID is simply a matter of inserting a record into a WAMS table that contains both the URI and the user ID. You can do this with the following class:

C#

```
[DataTable]
public class Registration {
    [DataMember]
    public int Id { get; set; }

    [DataMember]
    public string Uri { get; set; }

    [DataMember]
    public string UserId { get; set; }
}
```

VB.NET

```
<DataTable> _
Public Class Registration
    <DataMember> _
    Public Property Id As Integer

    <DataMember> _
    Public Property Uri As String

    <DataMember> _
    Public Property UserId As String
End Class
```

The existence of the ExpirationTime property on the PushNotificationChannel object suggests that channels might outlive the running of the current application. Therefore, it makes sense to have a strategy to refresh the channel.

The default expiration for a channel is 30 days. If an attempt is made to send a request to an expired channel, it will fail. It is up to the client application to manage and refresh the URI, so there is nothing that the service can do to request a new URI once the old one has expired. Therefore, the need to refresh the URI must also be part of your implementation.

The Mechanics of Push Notifications

When the application launches, it make a request to get the channel URI. Once you have the URI, log in to the provider of your choice and see if you have already registered a URI to the service. If so, make sure that the URI hasn't changed. If not, insert a new registration. The following code demonstrates how you can do this:

C#
```
PushNotificationChannel currentChannel;
MobileServiceUser currentUser = null;

MobileServiceUser currentUser;

private async void AcquirePushChannel() {
   currentChannel = await
    PushNotificationChannelManager.CreatePushNotificationChannelForApplicationAsync();
   var registrationTable = App.MobileService.GetTable<Registration>();
   var existingRegistration =
      registrationTable.Where(r => r.UserId == currentUser.UserId ).ToListAsync();

   if (existingRegistration.Count == 0)
      await registrationTable.InsertAsync(new Registration() {
         Uri = currentChannel.Uri,
         UserId = currentUser.UserId
      });
   else
      if (existingRegistration[0].Uri != currentChannel.Uri)
      {
         existingRegistration[0].Uri = currentChannel.Uri;
         await registrationTable.UpdateAsync(existingRegistration[0]);
      }
}
```

VB.NET
```
Dim currentChannel As PushNotificationChannel
Dim currentUser As MobileServiceUser

Private Async Sub AcquirePushChannel()
   currentChannel = Await _
    PushNotificationChannelManager.CreatePushNotificationChannelForApplicationAsync()

   Dim registrationTable = App.MobileService.GetTable(Of Registration)()
   Dim existingRegistration = _
      registrationTable.Where( _
      Function(r) r.UserId = currentUser.UserId).ToListAsync()

   If existingRegistration.Count = 0 Then
      Await registrationTable.InsertAsync(New Registration With { _
```

103

```
        .UserId = currentUser.UserId _
    })
 Else
    If existingRegistration(0).Uri != currentChannel.Uri Then
        existingRegistration(0).Uri = currentChannel.Uri
        Await registrationTable.UpdateAsync(existingRegistration(0))
    End If
  End If
End Sub
```

In the code, you can see how the `PushNotificationChannelManager` is used to generate the `PushNotificationChannel`. Then, using LINQ, a request is made to see if the user ID has already been registered. Finally, either the URI is updated (if the existing URI is different) or the registration is completed. At this point, the application is ready to process notifications.

SENDING NOTIFICATIONS

Notifications are sent through the server-side of WAMS. This means that the scripts related to the insertion, modification, or deletion of the records in one of your mobile services tables will be modified. Exactly which table and which scripts depend entirely on your application, but in each case, the mechanism used is the same, so the following examples are for an insertion. You can easily extend the sample code to the other scripts.

```
function insert(item, user, request) {
    request.execute({
        success: function() {
            request.respond();
            push.wns.sendToastText04(item.Uri, {
                text1: "The toast message goes here"
            }, {
                success: function(pushResponse) {
                    console.log("The update push was sent: ", pushResponse);
                }
            });
        }
    });
}
```

This script is a combination of functionality that is covered in earlier chapters, along with a new object (`push.wns`). The flow of logic starts with completing the insertion by using the execute method on the `request` object. If the insertion is successful, use the respond method to send a response back to the client. A side-effect of sending the response is to have the remainder of the script run in the background, without delaying the client any longer than necessary.

Sending Notifications

The heart of the push notification process is the push.wns object. A number (actually, a surprisingly large number) of methods are used to send notifications. The difference between the methods relates to the target (either a tile, a toast message, or a badge) and the format of the resulting message. For example, the sendTileWideText01 method sends a tile that looks like the one shown in Figure 6-8.

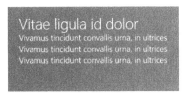

FIGURE 6-8

This format is fairly basic. The possible options include a collection of images, combination of images and text, or text that is in multiple columns, just for sending tiles. If you were to call sendToastImageAndText02, you would get a toast notification that looks like Figure 6-9.

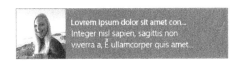

FIGURE 6-9

You get the idea. The options and parameters used by each of the methods vary based on the goal of the method. The one thing in common is that the first parameter is the URI for the channel. This is the URI that was recorded in the Registration table. Again, for simplicity, the preceding example presumes that the URI is part of the record being inserted. However, you can easily retrieve the URI from the Registration table using the mssql object that was demonstrated in Chapter 3.

These examples are just for pushing notifications through WNS. Corresponding push.gcm and push.apns objects send notifications through the Google and Apple networks as well. As an additional caveat, depending on the platform for the client, you might need to enable the receipt of toast notifications for your application. For instance, in a Windows Store application, the capability to receive toast messages must be enabled in the manifest. If you find that your application isn't receiving toast messages, an incorrect configuration should be one of the first places you look.

SUMMARY

Push notifications are an advanced feature made available through a combination of services that must work in a coordinated manner. Your mobile service needs to talk to either the Windows, Google or Apple notification services. Your client application needs to register a channel with the mobile service. And all of this needs to happen in the correct order so that notifications will arrive at their intended destination at the appropriate time. It can be complex to set up, but the functionality that is enabled can make a significant difference to the user experience.

The next chapter takes this idea to the next level. While earlier chapters discussed some of the scripting capabilities, Chapter 7 looks at a number of advanced features that can be used to enhance your application's functionality.

7
Advanced Scripting

IN THIS CHAPTER:

➤ Learn how to extend your class relationships into WAMS tables

➤ Understand how complex data types can be mapped into WAMS

➤ Automate your mobile service functionality by scheduling tasks in WAMS

WROX.COM CODE DOWNLOADS FOR THIS CHAPTER

The wrox.com code downloads for this chapter are found at http://www.wrox.com/go/windowsazuremobileservices on the Download Code tab. The code is in the Chapter 7 download and individually named according to the names throughout the chapter.

To this point, you have learned the basic (and not so basic) functionality offered within Windows Azure Mobile Services (WAMS). As a service that enables you to provide easily accessible data storage functionality, it is without peer. However, in addition to data creation, validation, authentication, and scripting, you still have a lot of areas to explore. In this chapter, you examine a number of different (and hopefully quite useful) scenarios. The idea is that by exploring complex solutions, you will not only learn to push the boundaries of what you can do using WAMS, but also get a solid sense of what is easily within the reach of your mobile service.

AUDITING UPDATES

Functionally, the idea of auditing a record is relatively commonplace. As part of the auditing process, a separate table is created that stores the details about any additions, modifications and deletions. For the example later in this section, you must have a table named audit in your mobile service. The attributes for the table are shown in Figure 7-1.

You can see that the table has five columns, with the following purposes:

- **id** — The sequential, numeric identifier for the record as required by WAMS
- **tableName** — The name of the table in which the record is modified
- **recordId** — The identifier for the record that is being modified
- **modificationDate** — The date and time on which the change was made
- **updateValues** — A JSON representation of the values that were updated

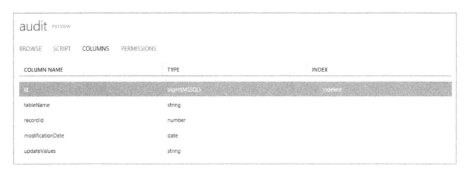

FIGURE 7-1

In order to participate in the auditing process, the client application doesn't need to do anything differently. Records are created, updated, and deleted using the InsertAsync, UpdateAsync, and DeleteAsync methods, as usual. And, by the way, this is the goal of any auditing mechanism: that the application updating the data isn't aware that the auditing is taking place.

The auditing mechanism is in the server-side script. Consider the following code, which is the update script for the table that is being audited:

```
function update(item, user, request) {
   request.execute({ success: addAuditRecord });

   function addAuditRecord() {
      var auditTable = tables.getTable('audit');
      var audit = {
         tableName: 'gameResult',
         recordId: item.id,
         modificationDate: new Date(),
         updateValues: JSON.stringify(item)
      };

      auditTable.insert(audit, {
         success: function() {
            request.respond();
         },
```

```
            error: function(err) {
                console.warn('An error creating an audit record', err)
                request.respond();
            }
        });
    }
}
```

The first step in the script is to execute the update. Upon the successful updating of the record, the auditing functionality is performed. As an approach, this is appropriate in that the auditing function can't get in the way of the update. If there were an issue with the creation of the audit record, the original record would still be updated.

Once the update has been performed successfully, the addAuditRecord function is called. This function retrieves the audit table using the getTable method. It creates a JavaScript object that represents the audit information, and then it inserts the record into the audit table using the insert method. Once all of these steps have been performed, a response is sent back to the client using the request.respond() method.

The only other component to be aware of in this script is the JSON.stringify method. This method is a part of JSON, which is a subset of JavaScript. Because the server-side code in WAMS is running in a node.js environment, all of the fundamental JavaScript components are available. In this case, the stringify method is used to convert a JSON object into a string representation.

Just to close the loop on this scenario, Figure 7-2 illustrates a value that has been inserted into the audit table as the result of updating a record in the gameResult table.

FIGURE 7-2

SUPPORTING OTHER DATA TYPES

WAMS provides native support for only a limited number of data types: Boolean, dates, numbers, and strings. Although these map onto more than four .NET data types, the reality is that being limited to these basic types can be restricting. However, it is possible to provide support for a wide range of different date types, including some of the more complex objects.

A relatively straightforward data type is TimeSpan. To be clear, the mechanism that is used supports any data type that is capable of being serialized and deserialized using a string value. So not only do TimeSpans work, but so could enumerations, URIs, or other classes with this capability.

The approach for simple unsupported data types is to add a converter that takes the object, and converts it to a string representation of a JSON object. This object can then be sent over the wire to the mobile service, where it is stored as a string. When the value is retrieved from the mobile service, it is returned into the client as a "stringified" JSON object, but the converter uses the values to create an instance of the original object.

The engine behind this process is represented in the IDataMemberJsonConverter interface:

C#

```
public interface IDataMemberJsonConverter
{
    object ConvertFromJson(IJsonValue value);
    IJsonValue ConvertToJson(object instance);
}
```

VB.NET

```
Public Interface IDataMemberJsonConverter

    Function ConvertFromJson(value As IJsonValue) As Object
    Function ConvertToJson(instance As Object) As IJsonValue

End Interface
```

The purpose of the two methods is fairly self-explanatory. The first one (ConvertFromJson) converts a JSON string into an object. The second (ConvertToJson) converts the values of an object into a JSON string. Consider the following implementation of this interface for a TimeSpan object:

C#

```
public class TimeSpanConverter : IDataMemberJsonConverter
{
    private static readonly IJsonValue NullJson = JsonValue.Parse("null");

    public object ConvertFromJson(IJsonValue value)
    {
        TimeSpan result = default(TimeSpan);
        if (value != null && value.ValueType == JsonValueType.String)
        {
            result = TimeSpan.Parse(value.GetString());
        }

        return result;
    }

    public IJsonValue ConvertToJson(object instance)
    {
        if (instance != null && instance is TimeSpan)
```

Supporting Other Data Types

```
        {
            return JsonValue.CreateStringValue(((TimeSpan)instance).ToString());
        }
        else
        {
            return NullJson;
        }
    }
}
```

VB.NET

```
Public Class TimeSpanConverter
    Implements IDataMemberJsonConverter

    Private Shared ReadOnly NullJson As IJsonValue _
        = JsonValue.Parse("null")

    Public Function ConvertFromJson(value As IJsonValue) As Object
        Dim result As TimeSpan = default(TimeSpan)
        If value Is Not Nothing And value.ValueType = JsonValueType.String Then
            result = TimeSpan.Parse(value.GetString());
        End If

        Return result
    End Function

    Public Function ConvertToJson(instance As Object) As IJsonValue
        If instance Is Not Nothing And instance is TimeSpan Then
            Return JsonValue.CreateStringValue(CType(instance, TimeSpan).ToString())
        Else
            Return NullJson
        End If
    End Function

End Class
```

Since the `TimeSpan` object is nullable, a static, read-only value is set up to represent a null JSON object. This value is returned if the object to convert has not been initialized. The default `TimeSpan` is set up in the `ConvertFromJson` method. If the incoming JSON object is not null and the type of object represented by the JSON object is a string, the `Parse` method on `TimeSpan` is used to convert the JSON object's string value into a `TimeSpan` object.

In the `ConvertToJson` method, the incoming object is checked to see if it is a `TimeSpan`. If so, the `CreateStringValue` method is used to convert the string representation of `TimeSpan` into a JSON object.

Now that you have a class for performing the conversion, you can look at how to attach the conversion code onto a method. Consider the following class declaration:

C#
```
public class GameClip
{
   public DateTime StartDate { get; set; }
   public TimeSpan Duration { get; set; }
}
```
VB.NET
```
Public Class GameClip

   Public StartDate As DateTime
   Public Duration As TimeSpan

End Class
```

Notice the property that has the type of TimeSpan. Just so you might recognize the problem were it to happen, an attempt to serialize the Duration object will result in an exception being thrown. Specifically, it's an ArgumentException claiming that an object of type TimeSpan cannot be serialized.

The final step is to mark the Duration property so that it uses the conversion class that you created. Modify the class declaration so that it looks like the following:

C#
```
public class GameClip
{
   public DateTime StartDate { get; set; }
   [DataMemberJsonConverter(ConverterType=typeof(TimeSpanConverter))]
   public TimeSpan Duration { get; set; }
}
```
VB.NET
```
Public Class GameClip

   public StartDate As DateTime
   <DataMemberJsonConverter(ConverterType := TypeOf(TimeSpanConverter))> _
   Public Duration As TimeSpan

End Class
```

The DataMemberJsonConvert attribute has a ConverterType property. That property is assigned the type that is used to convert the TimeSpan to and from JSON.

Like the auditing functionality described earlier in the chapter, this conversion doesn't require any assistance from the "other" side. Here, the conversion occurs on the client, and the server only sees the value as a string representation. The client has the capability to convert that string back into the corresponding .NET object when the item is retrieved from the mobile service.

SUPPORTING ARRAYS

The next level of complexity that you might find yourself facing is when the object that you're trying to save in your mobile service contains an array. Now you are faced with two possible implementations. First, you could simply serialize the array object into a JSON array and then store the string representation in the service. The downside to this approach is that you won't be able to perform any querying using the values in the array.

Just to illustrate this problem, consider a class that is declared as follows:

C#

```
public class Tournament
{
   public int Id { get; set; }
   public DateTime StartDate { get; set; }
   public DateTime EndDate { get; set; }
   public List<GameScore> GameScores { get; set; }
}
```

VB.NET

```
Public Class Tournament

    Public Id As Integer
    Public StartDate As DateTime
    Public EndDate As DateTime
    Public GameScores As List(Of GameScore)

End Class
```

Here, the `GameScore` class is defined as it appeared in Chapter 2:

C#

```
[DataTable(Name="gameResult")]
public class GameScore {
   [DataMember(Name="id")]
   public int Id { get; set; }

   [DataMember(Name="score")]
   public int Score { get; set; }

   [DataMember(Name="resultDate")]
   public DateTime GameDate { get; set; }
}
```

VB.NET

```
<DataTable(Name="gameResult")> _
Public Class GameScore
```

```
    <DataMember(Name="id")> _
    Public Property Id As Integer

    <DataMember(Name="score")> _
    Public Property Score As Integer

    <DataMember(Name="resultDate")> _
    Public Property GameDate As DateTime
End Class
```

If you were to create a JSON converter for the collection of `GameScore` objects and add the attribute to the `GameScores` property, the list of `GameScore` objects would be serialized as a string, and that would be just fine. However, if you wanted to find the highest score in a tournament, you would need to deserialize the entire list into local memory before the search could begin.

What if there was another way? Perhaps instead of storing the data as a string in the mobile service, the data could be put into a different table in the mobile service. But that's not something that you would like the client portion of the application to be aware of; the client should use the object as it is right now. How the data gets stored is an implementation detail of which the application should not be aware.

The key to addressing this scenario is to have an understanding of how the server-side pipeline works in WAMS. Figure 7-3 illustrates the different steps on the server side of WAMS.

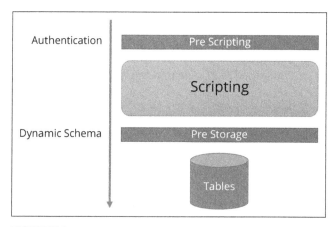

FIGURE 7-3

Before any of the scripts are executed, any authentication required is performed, and the payload is validated to ensure that it is reasonable. Keep in mind that the definition of "reasonable" here is just whether the request contains a valid JSON object. Once this pre-script processing has been performed, your scripts are executed. Those scripts can change the payload (by adding properties, for example) or can invoke other functionality (such as insertions into other tables).

Supporting Arrays

On the other side of the script execution, a pre-storage process ensures that the data put into the tables is valid. This step also dynamically creates the schema, presuming it has been enabled.

With this flow in mind, consider how you could handle arrays in WAMS. The idea is to ensure that the property in the object that you're sending to WAMS is serialized as a JSON array. Then in the scripting component, you can process the elements of that array into a different table.

To start with, you still need to have a converter on the GameScores property. For that data to be sent to the service, it needs to be serialized into an array of JSON objects:

C#
```csharp
public class GameScoreListConverter : IDataMemberJsonConverter
{
    private static readonly IJsonValue NullJson = JsonValue.Parse("null");

    public object ConvertFromJson(IJsonValue value)
    {
        throw new NotImplementedException();
    }

    public IJsonValue ConvertToJson(object instance)
    {
        List<GameScore> scores = instance as List<GameScore>;

        if (scores != null)
        {
            JsonArray arr = new JsonArray();
            foreach (var score in scores)
            {
                arr.Add(MobileServiceTableSerializer.Serialize(scores));
            }

            return arr;
        }
        else
        {
            return NullJson;
        }
    }
}
```

VB.NET
```vbnet
Public Class GameScoreListConverter
    Implements IDataMemberJsonConverter

    Private Shared ReadOnly NullJson As IJsonValue = JsonValue.Parse("null")

    Public Function ConvertFromJson(value As IJsonValue) As Object
```

```
        Throw New NotImplementedException()

    End Function

    Public Function ConvertToJson(instance As Object) As IJsonValue

        Dim scores As List(Of GameScore) = TryCast(instance, List(Of GameScore))

        If scores Is Not Nothing Then

            Dim arr As New JsonArray()
            Dim score As GameScore
            For Each score In scores
                arr.Add(MobileServiceTableSerializer.Serialize(scores))
            Next score

            Return arr
        End If
        Else
            Return NullJson
        End If

    End Function

End Class
```

This convertor looks (and acts) quite similar to the converter that was used for the `TimeSpan` class. The biggest difference is in how the information is moved back and forth from JSON. Here, you manually create a `JsonArray` object, and then populate it with the JSON serialized values from the list of `GameScore` objects. The result of the serialization (and what will go across the wire to your mobile service) is the following snippet:

```
GameScores: [
    {
        id: 10,
        score: 163,
        gameDate: '2013-02-16T05:57:21.242+00:00'
    },
    {
        id: 23,
        score: 223,
        gameDate: '2013-02-18T03:21:51.879+00:00'
    }
]
```

With this JSON in mind, one of the server-side scripts that will process this information is as follows:

```
function insert(item, user, request) {
    var scores = item.GameScores;
    item.GameScores = null;

    request.execute({ success: function() {
        var resultsTable = tables.getTable('gameResult');
        var count = scores.length;
        scores.forEach(function(score, index) {
            score.tournamentId = item.id;
            resultsTable.insert(score, {
                success: function() {
                    count++;
                    if (scores.length === count) {
                        request.respond();
                    }
                },
                error: function(err) {
                    console.warn('Error while inserting GameScore objects', err);
                    count++;
                    if (scores.length === count) {
                        request.respond();
                    }
                }
            });
        }});
    }
}
```

The first step in this script is to make a copy of the GameScores array, and set the value in the item being saved to null. There is no need to save a serialized version of the array in the Tournament table; setting the value to null will prevent that from happening. However, you do need the objects in the collection in order to save them in the gameResult table.

Next, the insertion request is executed. Once that has finished successfully, it's time to update the gameResult table. To do this, iterate across each of the GameScore objects copied at the beginning of the method. For each one, add the ID for that just-inserted tournament as a property. This allows the relationship to the parent record to be maintained. Then insert the object into the gameResult table.

You might have noticed the presence of a count variable. The Insert statement is called asynchronously. As a result, it is quite likely that the loop submitting the GameScore objects will have finished before the updates have actually taken place. So if a call to request.respond() was placed at the end of the method, the client would see a successful response before it was known that everything was successful.

By initializing the count variable to 0 and then incrementing it as each item is processed, the script can determine when the last of the requests have been completed. As soon as every one of the child records is inserted, the client application will be notified.

ADDITIONAL USER INFORMATION

Chapter 4 discussed the different options for authenticating a user. As it turns out, your mobile service can use Microsoft Account, Twitter, Facebook, or Google to validate who is trying to access your data. However, that mechanism only exposes the user ID. Each of the providers has the capability to expose additional data, but the standard authentication process doesn't present that information for you to use.

Fortunately, you're not stuck with this situation. With some creative scripting, you can get additional details about the user and take advantage of that within your application.

The key to accomplishing this is the User object that is passed into each of the script functions, along with the getIdentities method that the object exposes. The getIdentities method returns a value that depends on the provider used for authentication. For all four of the providers, the user ID and an access token is returned. However, if you are using Twitter as the provider, the result includes both an access token and a secret. Regardless of the details, it is possible to connect to the provider and extract additional information.

Facebook

Facebook exposes additional information through either an API library or via REST calls to the Facebook graph. If you're making calls from a server-side script, using REST is the simpler choice. The URL that is the endpoint for the graph is https://graph.facebook.com/*provider*, where *provider* is one of the supported endpoints. To gain access to the current user, you need to provide the access token as a query string parameter. The following script demonstrates how this is done:

```
function insert(item, user, request) {
    item.UserName = "who?";
    var identities = user.getIdentities();
    var req = require('request');
    if (identities.facebook) {
        var accessToken = identities.facebook.accessToken;
        var url = 'https://graph.facebook.com/me?access_token=' + accessToken;
        req(url, function (err, resp, body) {
            if (err || resp.statusCode !== 200) {
                console.error('Error sending data to Facebook: ', err);
                request.respond(statusCodes.INTERNAL_SERVER_ERROR, body);
            } else {
                try {
                    var userData = JSON.parse(body);
                    item.UserName = userData.name;
                    request.execute();
                } catch (ex) {
                    console.error('Error parsing response from Facebook: ', ex);
                    request.respond(statusCodes.INTERNAL_SERVER_ERROR, ex);
```

```
                }
            ]
        });
    } else {
        request.execute();
    }
}
```

At the top of the script, you see the call to `getIdentities` to retrieve the information for the currently authenticated user. The next line, the `require` statement, is one that has not yet been described.

WAMS provides a number of ready-made modules that can be included in the server-side scripts. These are incorporated into the script through the `require` function. In the preceding example, the `require('request')` call returns an object through which HTTP requests can be performed. Other modules that are available include `sendgrid` (which is used to send e-mails) and `azure` (which is used to manage Azure components).

In the remainder of the script, the URL to access the Facebook graph is constructed and submitted. If the request is successful, the body of the response is parsed as a JSON object and the `UserName` is extracted.

Google

The process for extracting additional information from Google is quite similar to Facebook. The difference is that the URL is (naturally) aimed at a Google endpoint, but the access token is passed as a query string parameter. Consider the following code as an example:

```
function insert(item, user, request) {
    item.UserName = "who?";
    var identities = user.getIdentities();
    var req = require('request');
    if (identities.google) {
        var accessToken = identities.google.accessToken;
        var url = 'https://www.googleapis.com/oauth2/v1/userinfo?access_token=' +
            accessToken;
        req(url, function (err, resp, body) {
            if (err || resp.statusCode !== 200) {
                console.error('Error sending data to Google: ', err);
                request.respond(statusCodes.INTERNAL_SERVER_ERROR, body);
            } else {
                try {
                    var userData = JSON.parse(body);
                    item.UserName = userData.name;
                    request.execute();
                } catch (ex) {
                    console.error('Error parsing response from Google: ', ex);
```

```
                    request.respond(statusCodes.INTERNAL_SERVER_ERROR, ex);
                }
            }
        });
    } else {
        request.execute();
    }
}
```

Microsoft Account

Accessing the additional information for a Microsoft Account follows the same steps as Google and Facebook. Again, the difference is the URI for the endpoint:

```
function insert(item, user, request) {
    item.UserName = "who?";
    var identities = user.getIdentities();
    var req = require('request');
    if (identities.microsoft) {
        var accessToken = identities.microsoft.accessToken;
        var url = 'https://apis.live.net/v5.0/me/?method=GET&access_token=' +
            accessToken;
        req(url, function (err, resp, body) {
            if (err || resp.statusCode !== 200) {
                console.error('Error sending data to Microsoft: ', err);
                request.respond(statusCodes.INTERNAL_SERVER_ERROR, body);
            } else {
                try {
                    var userData = JSON.parse(body);
                    item.UserName = userData.name;
                    request.execute();
                } catch (ex) {
                    console.error('Error parsing response from Microsoft: ', ex);
                    request.respond(statusCodes.INTERNAL_SERVER_ERROR, ex);
                }
            }
        });
    } else {
        request.execute();
    }
}
```

Twitter

The interaction with Twitter is similar, but slightly different. Instead of the access token, the UserId is provided on the URL. The actual Twitter user ID needs to be parsed out of the userId value that is returned from the identity. The following provides an example:

```
function insert(item, user, request) {
    item.UserName = "who?",
    var identities = user.getIdentities();
    var req = require('request');
    if (identities.twitter) {
        var userId = user.userId;
        var twitterId = userId.substring(userId.indexOf(':') + 1);
        url = 'https://api.twitter.com/users/' + twitterId;
        req(url, function (err, resp, body) {
            if (err || resp.statusCode !== 200) {
                console.error('Error sending data to Twitter: ', err);
                request.respond(statusCodes.INTERNAL_SERVER_ERROR, body);
            } else {
                try {
                    var userData = JSON.parse(body);
                    item.UserName = userData.name;
                    request.execute();
                } catch (ex) {
                    console.error('Error parsing response from Twitter: ', ex);
                    request.respond(statusCodes.INTERNAL_SERVER_ERROR, ex);
                }
            }
        });
    } else {
        request.execute();
    }
}
```

The examples for each of these providers demonstrate how to retrieve the username from the provider. Naturally, more information than just the username is available to the script. The technique is the same; only the property that you access is different.

SCHEDULING TASKS

One of the points to note for all of the scripts that you have seen is that the trigger for activation comes from the client side. For the scripts to execute, a record has to be inserted, updated, deleted, or retrieved. Although this model fits some (or even most) applications, at times you might want to have a script execute on a regular schedule. This type of interaction would allow, for instance, notifications to be pushed to your application on a regular basis instead of when an external event happens.

One of the more recent (at least as of the writing of this book) additions to Azure Mobile Services is the capability to schedule jobs, and when a job is executed, a server-side script is run.

To start with, navigate through the Windows Azure Management Portal to your mobile service. On the page for your mobile service, click the Scheduler link at the top to reveal the page shown in Figure 7-4.

CHAPTER 7 ADVANCED SCRIPTING

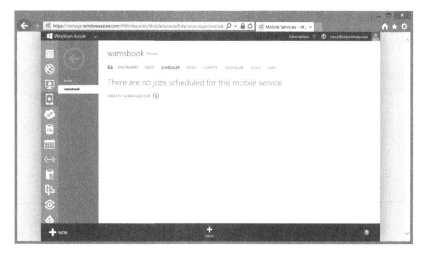

FIGURE 7-4

You can create a scheduled job by clicking either the Create icon at the bottom or the Create a Scheduled Job link near the top of the page. Whichever link you select, the dialog box shown in Figure 7-5 appears.

FIGURE 7-5

Provide a name for your job (the upcoming figures use `GetRSSUpdates`) and set a schedule. The interval can be in minutes, hours, days, or months. For an interval of minutes, the choice must be in increments of 15, meaning that scheduled jobs can't run more frequently than every 15 minutes. If you choose days, you specify not only the number of days in the interval, but also the starting date and time. The starting time becomes the time of the day that the job will run. For a month-based interval, the starting date and time are used to determine when subsequent jobs will run.

122

Scheduling Tasks

Once you have created the job, the screen shown in Figure 7-6 appears.

FIGURE 7-6

You'll notice that, by default, the job starts off as disabled. You can enable it by clicking the Enable icon at the bottom of the screen. However, at the moment there is no script functionality associated with the job, so you need to add some.

Click the scheduled job that you just created. The screen shown in Figure 7-7 appears.

FIGURE 7-7

Here, you can reconfigure the schedule for the job. In addition, after your job has been enabled and run a few times, the history for your job will be visible. To add functionality to your job, click the Script link, which displays the page shown in Figure 7-8.

FIGURE 7-8

Now you're in the same script editor that is used to create the table-based server-side scripts. You see a function that has the same name as your scheduled job. You put whatever functionality you would like to have executed on a regular basis inside this function. For instance, the following script will send toast notifications for the entries that are in the syndication feed for a blog. It presumes that there is a Channel table used to manage the registrations that have been made by various applications:

```
function GetRSSUpdates() {
var request = require('request');

request('http://blogs.objectsharp.com/syndication.axd',
     function postsLoaded (error, response, body) {
         var results = JSON.parse(body).results;
         if(results){
             results.forEach(function visitResult(post){
               sendNotifications(post);
             });
         }
     });
}
```

```
function sendNotifications(post){

var channelTable = tables.getTable('Channel');
channelTable.read({
    success: function(channels) {
        channels.forEach(function(channel) {

            push.wns.sendToastText01(channel.uri, {
                text1: post.title
            });
        });
    }
});
}
```

At this point, click the Save icon at the bottom of the page to save the script. You now have a job that is ready to be enabled. You can click the Run Once icon to test the script, and once it is ready to go, click the Enable icon to activate the scheduling functionality.

One final piece of information about scheduled tasks: If you are using one of the free Mobile Service instances, you are allowed only one scheduled job at a time. If you are running a reserved instance of WAMS, you can have up to 10 scheduled jobs.

SUMMARY

Windows Azure Mobile Services has the ability to support every level of functionality, from basic data access to sophisticated auditing, complex relationships and even scheduling background tasks. In other words, WAMS is a tool that you, as a developer, will enjoy having in your toolbox.

Which leads into our final chapter. Along with all of this functionality, there are capabilities and configurations that support common operational requirements. The support that WAMS provides to administrators and architects is discussed in Chapter 8.

8
Advanced Configuration

IN THIS CHAPTER:

➤ Scale your mobile service so that it can support a large number of users

➤ Utilize an existing database as the backing data store for your mobile service

➤ Integrate WAMS into your operations infrastructure

WROX.COM CODE DOWNLOADS FOR THIS CHAPTER

The wrox.com code downloads for this chapter are found at http://www.wrox.com/go/windowsazuremobileservices on the Download Code tab. The code is in the Chapter 8 download and individually named according to the names throughout the chapter.

Floating just under the surface of all that you have learned about Windows Azure Mobile Services (WAMS) to this point is a concept that might surprise you. At its heart, WAMS is just a combination of Windows Azure Web Sites and SQL Azure. Think of it like this: The endpoints that you communicate with using REST are services that could be constructed using WCF. Well, maybe not easily, but certainly feasibly. The data that is received, stored, and retrieved is persisted in a SQL Azure database. Could you create something that does what WAMS does? Of course you could. The challenge of building Windows Azure Mobile Services is not a technological one, nor are the benefits found in the technology.

What WAMS brings to the table is a combination of prebuilt functionality ready for you to take advantage of, along with the ability to extend its capabilities in a number of different directions. In Chapter 7, you looked at how to extend the functionality of WAMS through scripting. In this chapter, you look at the techniques used to enhance WAMS through configuration and back-end collaboration.

SCALING WAMS

The sample applications that demonstrate how to use Azure Mobile Services are nice, but they are exactly that…samples. By their very nature, they are intended to show one or two features, but are not necessarily indicative of what a commercial-quality app that uses WAMS extensively will look like. It is moving your application from "cute" to "real" that requires the most effort.

One of the areas of consideration as your application becomes more real is performance. However, because fixing the performance of your application is beyond the scope of this book, this chapter focuses on identifying issues that might exist within Azure Mobile Services itself — specifically, how to scale your mobile service and increase the robustness of the functionality that it provides.

In general terms, the goal of scaling is to increase the overall throughput of your mobile service. This is not to say that WAMS doesn't have the capability to support a decent number of concurrent users, even in the free version. But as the usage of your application increases (hopefully), so, too, will the load that is placed on the mobile service. At some point, the capacity of the service to process requests will be exceeded. This section discusses the options that are available to you once you reach that point.

Horizontal versus Vertical Scaling?

One of the first questions asked with respect to scaling is whether horizontal or vertical is the best choice. As is the case with many questions of this sort, the answer is "it depends." But before getting into that, take a look at Figure 8-1 for a basic description of the two styles of scaling.

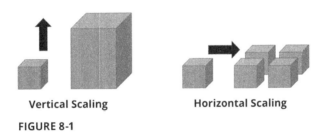

Vertical Scaling　　　**Horizontal Scaling**

FIGURE 8-1

On the left side of the figure is a visual representation of vertical scaling, or *scaling up*. The idea is to increase the power of the computing node without going outside of the node. Or, in more colloquial terms, scaling up involves increasing the horsepower of the computer. It could be more memory, faster CPU, or more core/CPUs, but the key concept is that the number of nodes does not increase.

On the right side of the figure is the visual representation of horizontal scaling, or *scaling out*. In opposition to a single compute node that characterizes vertical scaling, horizontal scaling adds compute nodes and then load balances between them. Instead of getting a bigger machine, you get more little ones.

There is no right or wrong when choosing the scaling approach you want to use. The biggest potential limitation is that, when scaling up, you can reach a practical limit for the capability of the machine — only so many cores and so much memory can be crammed into a single device.

Fortunately, when you are working with WAMS, you won't be approaching this limit. Although you have choices with regard to both scaling up and scaling out your mobile service, they are not unlimited in terms of capability. On the other hand, for the most part, your mobile service will be handling relatively straightforward REST-based requests. There is not a lot of processing power required, which means that, even with these limited choices, your ability to support a decently large number of users is well within the capacity of WAMS to satisfy.

To see what options are available, go to the Windows Azure Management Portal and navigate to your mobile service. In the links at the top of the dashboard, click Scale to get to the configuration screen for scaling your mobile service, shown in Figure 8-2.

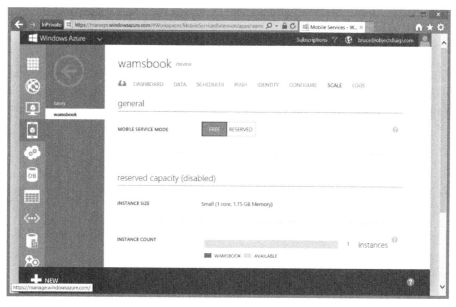

FIGURE 8-2

Three areas of configuration are available to you on this page. At the top, choose whether you are using the Free or Reserved version of Windows Azure Mobile Services. Two main differences exist between these versions. The first (and probably the most obvious) is the cost. Not too surprisingly, the Free version is free and the Reserved version is not. The details of how much it costs depend on how many instances you create (a process that is covered in a moment). Beyond the number of instances, given that the price Microsoft has been charging for the different components of Azure has been changing with great frequency, check out the current pricing by visiting http://www.windowsazure.com/en-us/pricing/calculator.

The second difference is less obvious. The free version of WAMS does have some limitations, such as limits in bandwidth, limits in CPU processing time, and limits in the amount of data that can be stored. The bandwidth limit is on a monthly basis. If you use more than a certain amount of bandwidth in a month, your mobile service will cease to function. The CPU limit is

Scaling WAMS

within an hour. If you use more than the allotted CPU within an hour, your mobile service will stop working. However, in an hour, the CPU will be reset and the service will start working again.

The data limits are more permanent. The free version of WAMS comes with 1 GB of data storage space. If you keep the data stored within that limit, there is no problem. If you need more space, you're going to pay for it.

The conversion of your mobile service from Free to Reserved is a simple one. Change the Mobile Service Mode (shown in Figure 8-2) to Reserved. Because you are going from a free mode in Windows Azure to one where payment might be involved, you will need to confirm this explicitly. A dialog box similar to the one shown in Figure 8-3 appears.

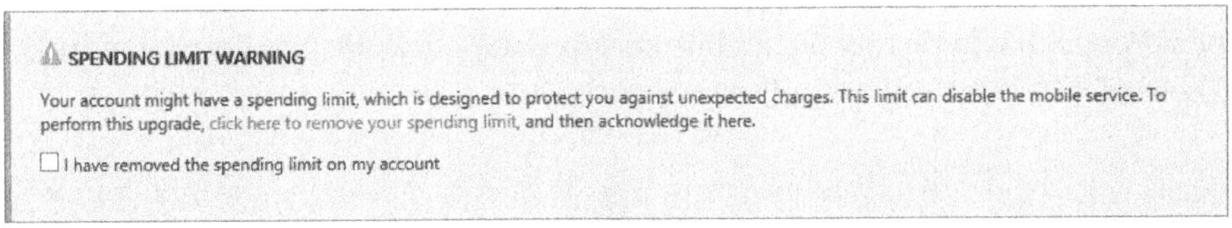

SPENDING LIMIT WARNING

Your account might have a spending limit, which is designed to protect you against unexpected charges. This limit can disable the mobile service. To perform this upgrade, click here to remove your spending limit, and then acknowledge it here.

☐ I have removed the spending limit on my account

FIGURE 8-3

If you haven't already done so, you need to remove the spending limit on your Azure account prior to converting to the Reserved version, or check the box that confirms that you've already removed the limit.

One final thing to be aware of is that once you convert a mobile service to Reserved, all of the services running in that same region will also be moved to Reserved. With all of the potential costs related to that move, this is something that you should think about as you create your mobile services. The limit to the number of free mobile services that you can have is 10 *per region*. If you expect to have some of your services in the Reserved mode and some in the Free mode, make sure that they are in different regions.

Once you have converted a service to Reserved, you have basically scaled up. The Free mode service is run in a shared environment, which means the machine on which your service runs is shared with other mobile services. When you are in Reserved mode, you have a machine that has been reserved exclusively for your service, which means you have increased the computing power that is available for your service to use.

However, being in Reserved mode gives you the ability to scale out as well. The machine that you are using is relatively low powered — it specs out to a single core with 1.75 GB of RAM. But you can also specify the number of instances of this single core machine that are allocated to your service. This is the scaling out capability of WAMS. You can add more instances of the single-core machine up to a total of 10.

The final portion of the page shown in Figure 8-2 relates to the size of the database that is the backing store for your service. You have two choices available to you: Web or Business. The default database is a Web edition with a size of 1 GB. However, by selecting the Business edition, you can arrange for a database to have up to 150 GB of storage space.

USING AN EXISTING DATABASE

The need for a database to act as the data store for WAMS has already been mentioned. When you create your mobile service, you have the ability to create a new database (which is the default, as well as what you did way back in Chapter 1). However, you can also use an existing database. That option is not only available when you create the mobile service, but also once your service is up and running.

To get to the point where you can utilize an existing database, get to the Windows Azure Management Portal, navigate to your mobile service, and click the Configuration link at the top of the dashboard. The page shown in Figure 8-4 appears.

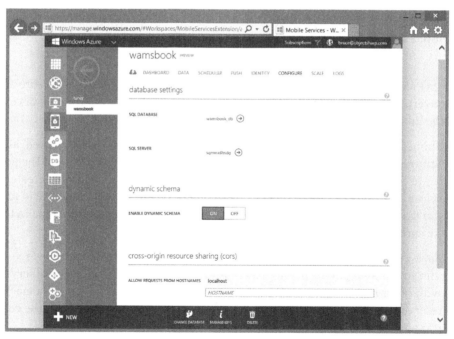

FIGURE 8-4

In the middle of the page, you can see there is a database and a database server associated with your mobile service. If you click the names or the arrow icons to the right of the names, you will go to the management screen for the database or the server. If you want to change the database that an existing mobile service is using, click the Change Database icon at the bottom of the page to reveal the dialog box shown in Figure 8-5.

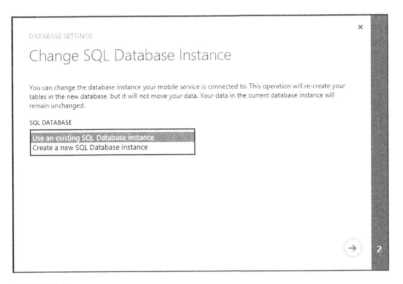

FIGURE 8-5

You can see from drop-down options that you have two choices: You can either create a new database, or connect to an existing one. Because this section is about attaching to an existing database, when you select that option and click the arrow at the bottom right of the dialog box, the second portion of the dialog box shown in Figure 8-6 appears.

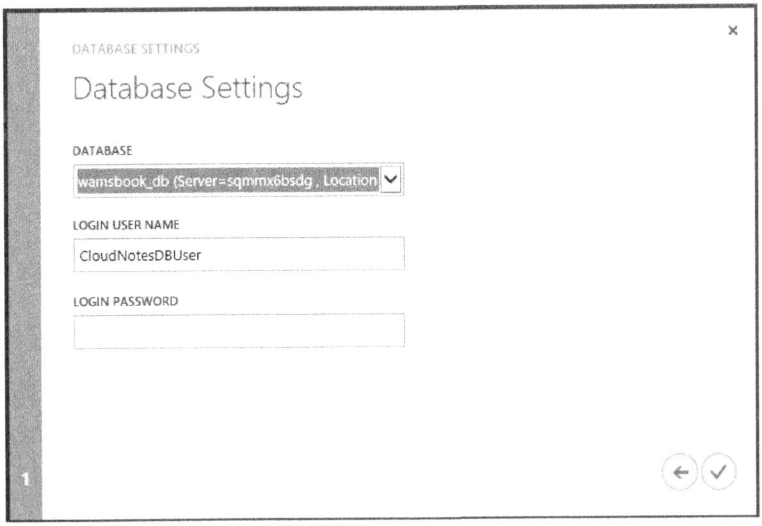

FIGURE 8-6

Here you select the database that you would like to use as the WAMS data store and provide the credentials for it. Once you have provided this information, click the checkmark icon at the bottom right to complete the database setup.

So that the second option is appropriately documented, if you create a new database by selecting the appropriate option in Figure 8-5, you are presented with the dialog box in Figure 8-7 so you can identify the database server and database name.

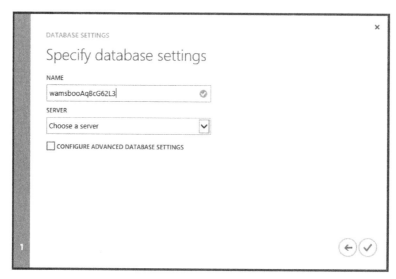

FIGURE 8-7

Setting up the database connection is not sufficient to allow your database to work with your mobile service. You must make some changes to your database to support working with WAMS. In a nutshell:

- The database tables need to be in a schema that has the same name as the mobile service
- The database tables need to have a clustered primary key named id with a type of integer.
- The tables in the mobile service need to be added manually after the database connection has been established

When WAMS is looking for the database table associated with a particular table in your mobile service, it uses a schema to scope the search. So if you have a mobile service with a name of WAMSBook, then the tables need to be in the WAMSBook schema. The one additional caveat to this is that if your mobile service has hyphens in the name (for example, WAMS-Book), they must be converted to underscores in the schema name (WAMS_Book).

The T-SQL statement to create a schema is the straightforward:

```
CREATE SCHEMA WAMSBook
```

Once you create the schema, you need to add each of the tables that are part of your mobile service to the schema using the following statement:

```
ALTER SCHEMA WAMSBook TRANSFER dbo.TableName
```

In this command, *WAMSBook* is the name of your mobile service and *dbo.TableName* is the fully qualified (that is, with the current schema, which is *dbo* in the snippet) of the table that you would like to be available through WAMS.

The second point mentions that there needs to be an integer column that is the primary key for the table. This is, at least at the moment, non-negotiable. The name of the column is also an unbendable requirement: the column must be named id. To make it even worse, the column name is case-sensitive, so make sure that you use lowercase letters. Add up all of these demands and the declaration for the column must be:

```
[id] [int] IDENTITY(1,1) NOT NULL
```

To be precise, you could use *bigint* instead of *int* and that too would work.

Finally, once the connection is ready to go and the tables are configured appropriately, you must add them manually to the mobile service. You do this by going to the Windows Azure Management Portal, navigating to your mobile service, clicking the Data option in the menu bar at the top, and then clicking the Create icon found at the bottom of the list of tables in your service. Make sure that the name of the table matches the table in the database and, once the table has been created, you will be able to access or update the data through the WAMS interface.

MONITORING WAMS

WAMS provides a mechanism that allows external systems to monitor the response time and update of your mobile service. To start, go to the Windows Azure Management Portal, navigate to your mobile service, and click the Configure option in the menu bar. As you scroll down the page, you will see a Monitoring section, as illustrated in Figure 8-8.

To be able to monitor your mobile service, you must be using a Reserved instance. The Free mode mobile service is not (at least, as of this writing) able to be monitored using this technique. In addition, this functionality is currently in preview mode. As a result, some aspects and features may be changed, removed, or added by the time this book goes into production.

At the moment, WAMS monitoring enables you to select up to three geographic locations from which to monitor the response and uptime of your mobile service. The standards used to determine acceptable performance are:

➤ Response time must be less than 30 seconds

➤ Error codes must be less than HTTP Status 400

FIGURE 8-8

To view the values for these metrics, click the Dashboard link in the top menu bar. WAMS checks on the response time and uptime for your service roughly every 15 minutes, and the results appear in the graph in the dashboard. Keep in mind that it might take a period of time (around 15–20 minutes according to the documentation) for the values to start appearing. But when they are active, you can see the response time and uptime in the line chart, as illustrated in Figure 8-9.

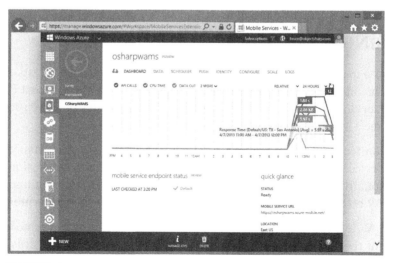

FIGURE 8-9

One final aspect of monitoring WAMS can be useful, especially as you write the various scripts. A console object is available through the scripts that you can use to write messages of various types to the log file. The console object itself implements four methods: `log`, `info`, `warn`, and `error`. All four of these methods have the same set of parameters. You pass the method a format string containing the text and value reference, and this string is followed by parameters that are referenced positionally within the string. The difference between these four methods is minimal. They simply change the logging level associated with the logged message. The `log` and `info` methods use the "info" logging-level. The `warn` method uses the "warning" log-level and the `error` method uses the "error" logging-level. This impacts how the messages appear in the mobile service, but doesn't change any other functionality of the methods.

To view logged messages, click the Log menu bar item in your mobile service. A screen similar to that shown in Figure 8-10 appears.

FIGURE 8-10

You can see the difference between the methods in the first column. The top message was emitted using the `log` method, whereas the last two came about through the `error` method.

Along with messages that you can create through your scripts, WAMS also injects its own messages into the log. If there is an error in one of your scripts, a log entry is made. In Figure 8-10, the second entry is because a script didn't have an appropriate exit point, and the third entry is because of a syntax error. To see the details (because the table doesn't show the entire message without some effort), select the desired entry and then click the Details icon at the bottom. A dialog box similar to Figure 8-11 appears.

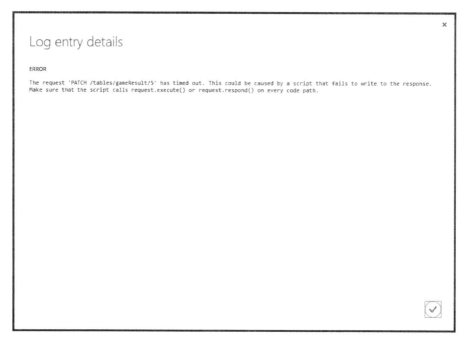

FIGURE 8-11

SUMMARY

As has been hopefully made clear throughout this book, Windows Azure Mobile Services is an interesting mix of simple and complex. The simplicity lies in the widely accessible (through a REST-based interface) data access functionality. The complexity is found in the server-side code, scheduled tasks, scalability, and operational configurability that is also available at a number of points with the services. This mix of simple and not-so-simple means that WAMS is actually appropriate for a wide variety of applications. And, ultimately, that is its goal.

But don't think that this is the 'final' product. As of this writing, WAMS is not officially out of beta yet. And if its brief history is any indication, you can expect to see more functionality and additional extensibility points being added as time goes on. Possibly even before it is out of beta. One of the benefits of working with an Internet-based service is that the barrier to upgrades and updates is low. And the Azure team at Microsoft seems to be taking advantage of this. So stay tuned. If you like what you see, then WAMS is already available for you. If it's not quite what you need, keep your eye on it. You never know what feature might have just been released.

INDEX

A

Access Control Services (ACS), 2, 7–8, 60
accounts
    developer accounts, 95–96
    Facebook, 61
    Google, 64, 66, 77
    Live ID, 9, 58, 67
    Microsoft Account, 9, 58, 67–69, 120
    Twitter, 69
ACS. *See* Access Control Services
Activate link, 10
Active Directory, Azure, 7–8
Add Class dialog box, 33
Add Column icon, 26
add reference, 28, 32
Add Score button, 34, 36
advanced configuration. *See* Mobile Services
advanced scripting. *See* scripting
aliveness, 93, 94, 95
Anybody with the Application Key, permission, 23
App class, 28, 34
App Secret, 63
AppFabric, 2
Apple Push Notification Services, 94, 99, 100
application key, 23, 26–27, 28, 88–89
App.xaml code file, 26, 28, 34
ArgumentException, 112
arrays
    advanced scripting, 113–117
    JavaScript, 56
    JSON, 54, 55, 113, 115
assembly, Windows Azure Mobile Services Managed Client, 28, 32
async keyword, 34
atomic transaction, 51, 55, 56, 57
auditing updates, 107–109
authentication, 58–78
    Azure Active Directory, 7–8
    client side, 74–77
    configure service, 72–74
    Facebook
        additional user information, scripting, 118–119
        authentication provider setup, 61–64
    federated, 58–61
    Flickr, 59
    Google
        additional user information, scripting, 119–120
        authentication provider setup, 64–67
    Microsoft Account
        additional user information, 120
        authentication provider setup, 67–69
    OAuth, 8, 61, 65
    REST, 88–92
    troubleshooting, 77
    Twitter
        additional user information, scripting, 120–121
        authentication provider setup, 69–72
authentication key, 48
await keyword, 34
Azure homepage, 9
Azure platform. *See also* Management Portal; Mobile Services
    Access Control Services, 2, 7–8, 60
    Active Directory, 7–8

Cloud Services, 4–5, 7, 8
components, 3–8
connectivity options
    Azure Connect, 2–3, 7
    Azure Virtual Network, 6–7
    Traffic Manager, 7
Data Storage
    Blobs, 6
    choose storage solution, 5
    SQL Database service, 5, 11, 23
    Tables, 5–6
execution models, 3–5
history, 2–3
IaaS, 3, 4
introduction, 1–2
PaaS, 3
relational database in cloud, 2
SaaS, 3
Virtual Machines, 1, 2, 3, 4, 5, 6
virtualization solutions, 1–2
Web Sites, 4, 126

B

binary notifications, 95
Bit, 19
Blobs, Azure, 6
Boolean, 19, 31, 42, 86, 109
btnUpdateScore_Click, 35, 36

C

Callback URL, 70
CAPTCHA, 71
Certificate Signing Request, 99
changeGameDate, 53
Choose Platform section, 14
Click event handlers, 34–36, 74
Client ID, 64, 65, 67, 69, 72
Client Secret, 39, 67, 72, 98, 100
client-side authentication, 74–77
client-side functionality, 26–38
client-side libraries, 79
cloud computing

    execution models, 3–5
    technology impact, 1
Cloud Services, Azure, 4–5, 7, 8
CLR data types, 19
columns, 20–26
complexity-simplicity, WAMS, 136
computer, network as, 1
configuration. *See* Mobile Services
Configure Server Key for API Project dialog
    box, 98–99
connectivity options
    Azure Connect, 2–3, 7
    Azure Virtual Network, 6–7
    Traffic Manager, 7
Consumer Technology Preview, 2
count variable, 117
Create, 49
Create a New Windows Store App link, 14
Create an Application page, 69–70
Create an OAuth 2.0 Client ID button, 65
Create Client ID button, 67
Create New App button, 63
Create New Table dialog box, 21–22
Create Your Twitter Application, 71
CreateCustomer, 79
createPushNotificationChannelFor
    ApplicationAsync(), 101, 103
createPushNotificationChannel
    ForSecondaryTileAsync, 101
CRUD operations
    client-side functionality, 28–36
    common scenarios
        inserting multiple items per call,
            54–57
        storing per-user data, 48–51
    server-side scripting
        del function, 42, 44–45, 51
        insert function, 42–43
        retrieve function, 45–47, 52
        update function, 43–44
    uniform interface, REST, 79–80
currentScore variable, 34–35, 36
Customer Key, 72
Customer secret values, 72

D

data centers, Traffic Manager and, 7
data creation and manipulation, 17–38
Data link, 40, 73
data model, 18–26
data persistence function, SQL table, 14
data storage
    Azure Data Storage
        Blobs, 6
        choose storage solution, 5
        SQL Database service, 5, 11, 23
        Tables, 5–6
    Mobile Services functionality, 17, 21, 38
data transfer class, 17, 21, 49
Data Transfer Object (DTO), 20–21
data types
    Boolean, 19, 31, 42, 86, 109
    CLR, 19
    float, 19
    floating point, 19
    integer, 19, 21, 132
    JSON, 19
    .NET, 109
    nvarchar(max), 19, 53
    support, 109–112
    TimeSpan, 109–112, 116
    T-SQL, 19
Database dropdown, New Mobile Service dialog, 11
databases
    SQL Database service, 5, 11, 23
    use existing database, 11, 130–133
DataContractJsonSerializer, 84
DataMember attribute, 20–21, 85
DataMemberJsonConverter, 54, 55, 112
DataTable attribute, 20–21, 29, 54
dates
    JSON, 53
    UTC-formatted, 84
DateTime, 19
DateTimeOffset, 19
decimal, 19
del function, 42, 44–45, 51
Delete, 21, 49

DELETE, 81–82, 83
Delete Column icon, 26
Delete Score button, 36
DeleteAsync( ), 30, 36, 108
deleting record, 30
developer accounts, 95–96
Developer License, 14
Developer Program, iOS, 99
dialog box
    Add Class, 33
    Configure Server Key for API Project, 98–99
    Create New App, 63
    Create New Table, 21–22
    New Mobile Service, 10–12
    Reference Manager, 32
Drupal, 4
DTO. *See* Data Transfer Object
Duration property, 112
dynamic schema generation, 18–20, 23, 49, 53, 114

E

Enable Dynamic Schema, 20
enumerations, 75, 109
error handling, 37–38
event handlers, Click, 34–36, 74
Everyone, permission, 23
exceptions
    InvalidOperationException, 38, 74, 75, 77
    MobileServiceInvalidOperation Exception, 38
execute method, 42, 44, 55, 104
execution models, Azure platform, 3–5
ExpirationTime property, 101, 102
extensibility, 126, 136. *See also* scripting

F

Facebook
    additional user information, scripting, 118–119
    authentication provider setup, 61–64
federated authentication, 58–61
Fiddler, 77

FIFO, 8
$filter, 86
Flickr, 59
float, 19
floating point, 19
400 status code, 51, 134
403 status code, 43
404 status code, 50
free version, WAMS, 11, 127–129
full-trust, 2

G

gallery images, Azure Web Sites, 4
GameScore, 32–37
    arrays, 113–117
    columns, 18–26
    DataMember attribute, 20–21, 85
    DataTable attribute, 20–21, 29, 54
    dynamic schema generation, 18–20
    IEnumerable(Of GameScore), 31, 54, 55
    permissions, 21, 23, 26, 27
GET, 77, 80, 81, 82, 83, 84, 86, 87
GetCustomersByName, 79
GetCustomersByRegion, 79
GetTable generic method, 29, 31
getting application key, 26–27
Google
    additional user information, scripting, 119–120
    authentication provider setup, 64–67
Google account, 64, 66, 77
Google Cloud Messaging, 94, 98–99

H

horizontal scaling, 127–130
HTTP
    DELETE, 81–82, 83
    400 status code, 51, 134
    403 status code, 43
    404 status code, 50
    GET, 77, 80, 81, 82, 83, 84, 86, 87
    POST, 80–81, 83, 87, 88, 90
    PUT, 80, 81, 82, 83
    302 response, 77
HTTPWebRequest, 83
HTTPWebResponse, 83

I

IaaS. *See* Infrastructure as a Service
icon, Mobile Services, 9, 10, 21, 26
Id column, 26
IDataMemberJsonConverter, 54, 110, 111, 115
idempotency, 81
IEnumerable(Of GameScore), 31, 54, 55
IgnoreDataMember attribute, 21
IIS (Internet Information Services), 4, 99
IMobileServiceTable generic interface, 31
Imports statement, 32
Infrastructure as a Service (IaaS), 3, 4
$inlinecount, 86
insert function, 42–43
InsertAsync(), 29, 34, 103, 108
inserting
    multiple items per call, 54–57
    records, 28–29
insertNext(), 52, 53
InsertProduct, 79
integer, 19, 21, 132
Internet Information Services. *See* IIS
Invalid user message, 51
InvalidOperationException, 38, 74, 75, 77
iOS
    applications, 14
    Developer Program, 99
    devices, 94
    Provisioning Portal, 99
IP addresses, 6–7, 23–24

J

Java, 2
JavaScript. *See also* scripting
    array, 56
    functions, 45, 46, 47, 48, 50
    server-side scripting, 39, 42, 48

Index

`JavaScriptSerializer`, 84
Joomla, 4
JSON
    array, 54, 55, 113, 115
    data types, 19
    date, 53
    objects, dynamic schema generation, 18–20
Json.NET, 84–85, 87

K

keys
    Anybody with the Application Key, permission, 23
    application key, 23, 26–27, 28, 88–89
    authentication key, 48
    Customer Key, 72
    getting application key, 26–27
    key/value pairs, 6
    Manage Keys icon, 26–27, 91
    master key, 23, 27, 48, 91, 92
    shared key, 61
keywords, 34, 42

L

`lastUpdated`, 43, 44
libraries
    client-side, 79
    Facebook API, 118
    Json.NET, 84–85, 87
    MSDN Library documents, 14
LINQ, 31, 46, 104
Linux, 3, 4, 5
Live ID, 9, 58, 67. *See also* Microsoft Account
`LoginAsync`, 75, 76

M

`MainPage.xaml` file, 32–33, 34
Manage Keys icon, 26–27, 91
Manage NuGet Packages option, 84
Managed Client, Windows Azure Mobile Services, 28, 32

Management Portal, Azure
    Mobile Services setup, 9–13
    Portal link, 9
    table creation, 21
Management Studio, SQL, 5, 23
manipulation of data. *See* data creation and manipulation
master key, 23, 27, 48, 91, 92
Messaging
    Azure Queues, 8
    Service Bus, 2, 8
Microsoft Account, 9, 58, 67–69, 120
`Microsoft.WindowsAzure.MobileServices` namespace, 20
Mobile Services (Windows Azure Mobile Services-WAMS). *See also* REST; scripting
    Azure Web Sites compared to, 126
    complexity-simplicity, 136
    configuration
        introduction, 126
        monitoring WAMS, 133–136
        push notifications, 100
        scaling, 126–130
        use existing database, 11, 130–133
    described, 9
    extensibility, 126, 136
    free version, 11, 127–129
    fundamental function, 17, 21, 38
    introduction, 8–9
    purpose, 17, 21, 38
    Quick Start screen, 13–14, 21, 26
    Reserved mode, 128–129, 133
    REST API, 84–89
    setting up, 9–13
    simplicity-complexity, 136
    SQL Database compared to, 126
    strength of, 9
Mobile Services icon, 9, 10, 21, 26
Mobile Services SDK, 14, 17, 28, 74
`MobileService` read-only field, 28
`MobileServiceAuthenticationProvider`, 74, 75, 76
`MobileServiceInvalidOperation Exception`, 38

MobileServiceTableSerializer, 55, 115
MobileServiceUser, 74, 76, 103
model-first development, 21
modify existing record, 30–31
ModifyProduct, 79
monitoring Mobile Services, 133–136
MSDeploy, 3
MSDN Library documents, 14
multiple items per call, inserting, 54–57

N

Name property, 20, 21, 29, 54
namespace
    Microsoft.WindowsAzure.
      MobileServices, 20
    System.Runtime.Serialization, 20
naming standards, mobile services, 11
.NET data types, 109
network, as computer, 1
New Mobile Service dialog box, 10–12
Newton-King, James, 84
notifications. *See* push notifications
NuGet package, 84
nullipotent, 81
nvarchar(max), 19, 53

O

OAuth, 8, 61, 65
Objective-C, 26, 83
OData. *See* Open Data Protocol
one-to-many communication, 8
Only Authenticated Users, permission, 23
Only Scripts and Admins, permission, 23
OnNavigatedTo, 36
Open Data Protocol (OData), 83, 86, 87
$orderby, 86

P

PaaS. *See* Platform as a Service
PATCH, 82, 83, 87

payment model
    free version, WAMS, 11, 127–129
    Virtual Machines, 4
permissions, 21, 23, 26, 27, 47, 48, 60, 88
Permissions link, 73
per-user data, storing, 48–51
photo-printing service, 59–61
photo-sharing service, 59–61
PHP, 2, 83
Platform as a Service (PaaS), 3
polling request, 93. *See also* push notifications
Portal. *See* Management Portal
POST, 80–81, 83, 87, 88, 90
PowerShell scripts, 2
property bag, 6
Provisioning Portal, iOS, 99
proxy classes, 18, 76, 77, 84, 92
proxy object, 28
pull request, 93. *See also* push notifications
push notifications, 93–106
PushNotificationChannelManager, 101, 103, 104
push.wns, 104, 105
PUT, 80, 81, 82, 83

Q

Query object, 45, 46
query string parameters, REST API, 86
Queues, Azure, 8
Quick Start screen, 13–14, 21, 26

R

raw notifications, 95, 101
read method, 45, 50
records
    deleting, 30
    inserting, 28–29
    modify, 30–31
red squiggly underlines, 32
reference, 28, 32
Reference Manager dialog box, 32

Index

refreshList( ), 34, 35, 36
region, New Mobile Service dialog, 11
register application, push notifications, 95–99
relational database in cloud, 2
Remote Desktop Connection, 4
Representational State Transfer. *See* REST
requesting channel, push notifications, 101–104
Reserved mode, WAMS, 128–129, 133
respond method, 44, 104, 109
REST (Representational State Transfer), 79–92
    authentication, 88–92
    Mobile Services support, 82–88
    query string parameters, 86
    uniform interface, 79–80
    virtual machine creation, 4
    WAMS REST API, 84–89
Retrieve, 21, 31–32, 49
retrieve function, 45–47, 52

S

SaaS. *See* Software as a Service
sample applications, 13–15, 26–28, 126
scaling, 126–130
scaling out, 127–128, 129
scaling up, 127–128
scheduling tasks, 121–125
schemas
    dynamic schema generation, 18–20, 23, 49, 53, 114
    Enable Dynamic Schema, 20
    SQL, 5, 6
Score text box, 36
ScoreListConverter, 54, 55
Script link, 41, 124
scripting
    additional user information
        Facebook, 118–119
        Google, 119–120
        Microsoft Account, 120
        Twitter, 120–121

advanced scripting
    arrays, 113–117
    auditing updates, 107–109
    data types support, 109–112
    introduction, 107
    scheduled tasks, 121–125
PowerShell, 2
server-side scripting, 39–57
    accessing, 39–42
    del function, 42, 44–45, 51
    insert function, 42–43
    JavaScript, 39, 42, 48
    JavaScript functions, 45, 46, 47, 48, 50
    retrieve function, 45–47, 52
    update function, 43–44
    User object, 47–48, 49
SDK, Mobile Services, 14, 17, 28, 74
Secure Token Service. *See* STS
$select, 86
sending notifications, 104–105
server-side scripting. *See* scripting
Service Bus, 2, 8
service-based delineation, 3
shared key, 61
shared secret, 61
simplicity-complexity, WAMS, 136
single sign-on service, 8, 59
Site URL, 26, 27, 63
$skip, 86
Software as a Service (SaaS), 3
Solution Explorer, 32, 33
SQL Database service, 5, 11, 23
SQL schemas, 5, 6
SQL Server, 5, 6, 21, 56
SQL Server Management Studio, 5, 23
status code
    400, 51, 134
    403, 43
    404, 50
statusCodes.FORBIDDEN, 42, 44
statusCodes.OK, 44
statusCodes.UNAUTHORIZED, 44
storing per-user data, 48–51
strings, 19, 109

INDEX

STS (Secure Token Service), 59–61
    provider
        Facebook, 61–64
        Google, 64–67
        Microsoft Account, 67–69
        Twitter, 69–72
Subscription dropdown, New Mobile Service dialog, 11
`System.Runtime.Serialization` namespace, 20

T

tables. *See also* `GameScore`
    Azure Tables, 5–6
    columns, 20–26
    Create New Table dialog box, 21–22
    creation, Management Portal, 21
    permissions, 21, 23, 26, 27
    `TodoItem`, 13–15
tasks
    MSDN Library documents, 14
    scheduled, 121–125
technology, cloud's impact, 1
302, HTTP, 77
`TimeSpan`, 109–112, 116
toast notifications, 93, 95, 101, 104, 105, 124
`TodoItem` table, 13–15
tokens, Active Directory, 7. *See also* STS
`$top`, 86
Traffic Manager, 7
troubleshooting authentication, 77
trust. *See also* STS
    federated authentication, 58–61
    full-trust, 2
try-catch block, 37
T-SQL, 19, 56, 133
Twitter
    additional user information, scripting, 120–121
    authentication provider setup, 69–72

U

uniform interface, REST, 79–80
Update, 21, 49
update function, 43–44
`UpdateAsync( )`, 30, 36, 103, 104, 108
`UpdateCustomer`, 79
updating data, WAMS REST API, 87–88
use existing database, 11, 130–133
`User` object, 47–48, 49, 61, 118
`userId` property, 45, 47, 49, 61
`using` statement, 32
UTC-formatted dates, 84

V

validation, 39–57. *See also* scripting
    common scenarios
        inserting multiple items per call, 54–57
        storing per-user data, 48–51
    introduction, 39
    summary, 57
vertical scaling, 127–130
Virtual Machines, Azure, 1, 2, 3, 4, 5, 6
Virtual Network, Azure, 6–7
virtualization, 1–2
Visual Studio 2012, 14, 32, 84
Visual Studio Express, 14
VPN gateway device, 6, 7

W

WAMS. *See* Mobile Services
WCF. *See* Windows Communication Foundation
web applications, 2, 3, 4, 5, 39, 67, 80
Web roles, 2, 4–5, 8
Web Sites, Azure, 4, 126
WHERE clause, 47
where method, 31, 45, 46, 49, 50
Windows Azure Mobile Services Managed Client, 28, 32

Windows Communication Foundation (WCF), 3, 126
Windows Notification Services (WNS), 94, 95–98
Windows Phone 8 applications, 14, 32, 93, 95
Windows Store applications, 14, 32, 75, 95, 105
WNS. *See* Windows Notification Services
WordPress, 4
Worker roles, 2, 4, 5, 8

XAML
    App.xaml code file, 26, 28, 34
    authentication on client side, 74–75
    MainPage.xaml file, 32–33, 34
XML notifications, 95